(Working title and Subtitle)

Creation, Science, and the Scientific Theory of Evolution vs Evolutionism

(or: Science Six Feet Out on the Fog*)

by David Bump

Introduction

Why did I write a book about creationism and evolutionism, when there are so many others, including many written by creationist scientists and other professionals? Because none that I've seen gives an organized overview of the subject that advances steadily from start to finish, from the foundations of philosophy and ancient history to 21st-century discoveries and applications of science. This book is intended to be an introduction to the debate for those who haven't already gotten deeply into it, but would like to. I believe that many who have gotten deeply into debating the details also need to step back and look at the big picture. I start with the roots of the debate, and go through the history of faith, philosophy, and science behind beliefs about the origin and history of life on Earth.

As a young-Earth creationist, I don't pretend to be free of bias, but I hope I have set a good example for calm and reasonable study of the subject. I have tried to fairly represent a number of reasons and facts which have led others to believe in evolutionism and other old-Earth views. No complete set of beliefs regarding the origins of the universe, Solar System, Earth, life, and all the major forms of life, can be proven by logic and observation to everyone's satisfaction. We all have to have faith in something (or Someone) beyond the things we can observe or demonstrate to be true, or possible.

Knowing the ways we think about all these things is more important than knowing the facts and arguments presented by both sides. We might all get along a bit better if we keep in mind that we all have beliefs that we can't prove. Each of the four parts of this book will look at the philosophical roots of thinking in Western societies about origins.

One part of the debate that is often ignored, badly misunderstood, or little known, is its history. Starting with the beginning of religion and ideas about creation, to early evolutionary explanations, to the significance of Darwin's work and on to the historical effects of Darwinism, there are many things that are seldom taught, while other things are presented as facts that aren't true at all. A walk through history will give us a familiar chronological order to build on, and will be at the heart of each part of the book.

Each of the four parts will end with some of the differing explanations for the observations that everyone agrees on. The debate continues with such rancor because so much that we can agree on is downplayed, while opinions that depend heavily (by faith) on assumptions are considered facts. This book shows how these assumptions were smuggled into science, causing some areas of science to go beyond their reasonable bounds and creating conflict in areas where science was not designed to go. I will show some of the key examples of evidence for evolution, and explain why creationists feel they are compatible with divine creation, or are not reliable evidence.

There are many different disguises of beliefs that can make them appear to be facts. Everyone is aware that Creationism involves beliefs that can't be scientifically demonstrated, but it seems fewer realize that such beliefs underlie what many consider scientific facts. By showing the historical background and the basic philosophies behind the creation-evolution debate in this organized way, I hope to help people on both sides understand better why we have different explanations for the same facts. In some areas, we may have to realize that we see things so differently that debate is almost useless. I believe, however, that our understanding of the world would be greatly improved if scientists working within the creation framework were given full support and acceptance within the scientific community, rather than being shut out because of the monopoly of the naturalistic philosophy behind evolutionism.

*Note: "Six feet out on the fog"** is taken from Carl Sandburg's "Yarns of the People" (part of the larger work, *The People, Yes*) in which one of the tall tales he mentions is "a fog so thick we shingl'ed the barn and six feet out on the fog." In my search for the reason why the evolution-creation debate continues with people on both sides claiming that science clearly supports their side and rules out the other, I have concluded that the problem is a philosophical fog that became part of the official practice of some areas of science. Many people equate this philosophical view with science itself. Science as it was originally conceived, practiced, and taught was an attempt to avoid the endless debating of philosophy, and stick to ideas that can be tested with repeated observations and experimentation. Unfortunately, the ounce of philosophy that every human endeavor must include grew in the fertile soil of formal scientific establishments until it became the proverbial tail wagging the dog (aren't mixed metaphors FUN?). When creationists say that science disproves evolutionism, they rely on the original concept of science, but evolutionists use the modern, philosophical sense of "Science" and claim that creation cannot be part of it. They are both right, because the authorities currently defining what science is have embraced the fog of philosophical naturalism rather than keeping to the foundation of the reliable method of science that brought us the technological wonders we use every day.

Acknowledgments

I am not an expert or professional, so all I can ask is that you keep an open mind and investigate for yourself the line between what we truly know and what we only think we know. As far as I have gotten things right, I thank God first and foremost. There are many others I thank, such as my parents and the rest of my family, many individuals in the creation science community (especially those in the Creation Research Society with whom I've corresponded via e-mail), Karl Priest, teachers from kindergarten through college (Bob Jones University in Greenville, SC), and more. Although I don't want to clutter my writing with all the references to all the books that should be mentioned, of course most of what I know (or think I know) I learned from somebody else, and I will try to list some of the most relevant sources somewhere in each appropriate chapter. Here I would like to acknowledge for their general inspiration two of the works of Douglas R. Hofstadter, *Metamagical Themas: Questing for the Essence of Mind and Pattern* (Basic Books, New York 1985) and *Gödel, Escher, Bach: An Eternal Golden Braid* (Vintage Books, New York, 1980), which I read in that reverse-chronological order.

Map of the Journey

Part I: First Things First

Part II: A New Game Is Invented: Science

Part III: Faith in Reason and Nature...

Part IV: Dogmas – Incompatible Worldviews Fuel Endless Debate

Part I: First Things First

Chapter 1: Read the instructions before playing the game.

Step One: Step back and look in the mirror.

Very few people, if any, are properly prepared when they first run into the "creation-evolution debate." We are all brought up to believe one way or another on the subject. Few if any of us have been trained from our youth in logic, philosophy, and the details of the history of science and the differences in beliefs. Any honest person should admit that some things they once accepted as facts have turned out to be false, or at best merely opinions. It's more difficult to consider that what we now believe could also be wrong.

A major reason for this is the inability to understand why "they" can't see that "our" view is the right one. It all seems so obvious. There seem to be so many facts supporting what "we" believe. Whatever gaps, difficulties, or problems "we" discover or "they" bring up seem unimportant or easily explained away. At any rate, such problems don't seem nearly as bad as the faults in what "they" believe. This is the way we all tend to see the world. Perhaps if we keep in mind that their apparent willful stupidity is pretty much the same as our brilliant reasoning looks to them, we will all be a bit calmer and more understanding in our discussions, debates, and arguments.

Unfortunately, many arguments will never be settled by reasoned debate alone. It is tempting to wish that all humans were perfectly logical, in the belief that they would all see the logic of our position and stop arguing with us. On the other hand, as many science fiction tales have shown, there is much more to being human than our ability to reason logically, and we don't want to give up all of that. Besides, in the complexities of the real world, there may be no logical way to decide which logical principles to use, or which assumptions to begin with. There are some things that can't be decided upon logically, and yet they have to be decided before we try to discern the history of life. Different starting points, whose logical value cannot be decided by logic alone, lead to different but equally logical conclusions.

The significance of the previous paragraph is that both sides should realize first of all that the other side isn't stupid, they're just playing by a different set of rules. It isn't a question of not being able to see facts, but which facts we find most impressive, and the logical frameworks through which we see them. No matter how much data we share and compare, no matter how well we present our arguments and defenses, it will not mean the same thing to the other side. This does not mean that debating is pointless. Individuals have changed sides after studying the data, issues, and arguments. What it does mean is that we should all be calm and gracious, even when the other guys seem to be willfully ignoring the obvious.

So, almost all of us come to the debate thoroughly convinced of the truth of what we've been taught, although none of us has been taught in an organized, thorough, and error-free way. Our mental arsenals (on both sides) may be cluttered with vague recollections of grade school and Sunday School lessons, some catchy slogans, zingers, "stumpers," and perhaps "proofs" that have long ago been discarded by those who know better. If the shock of the first encounter with someone from the other side does not cause us to convert and become a more stubborn warrior for our new cause, it may cause us to become more entrenched in our beliefs, all the more convinced of the foolishness or evil of everyone on the other side.

For decades the dust of this battle has been especially thick, as both sides shout variations and refinements of the same slogans and arguments. Many seem to work very hard at studying whatever specific point they happen to be arguing at the time with the single-minded conviction that no sign of weakness must be shown, no concessions made to the other side. If they study what the other side has written themselves, rather than the mockeries presented by the cheerleaders for their own side, it is only with an eye to finding some flaw, some weak point to harp on.

I'd like to think that there are exceptions, and that I'm one them, having been surprised by the strength of some evolutionary arguments and disappointed with the weaknesses of some for Creation and a young Earth, yet neither being converted, nor becoming a wild-eyed berserker with eyes closed and verbal cudgel swinging.

I don't have any special credentials to impress you with. By the same token, I don't have a career in science to protect, nor a religious ministry to promote. I was raised as a Bible-believing Christian, but I have also always had a great appreciation for science. I have been reading scientific magazines and journals since I was a young child, including *National Geographic, Science News, Scientific American,* and *Nature.* I've had a few letters to the editors published. I've read Darwin's *Origin of the Species* and more recent works by evolutionists such as Sagan and Dawkins. Yes, I'm always looking for the flaws, but I am also looking for challenges to what I believe, too. I've also tried to understand belief in evolutionism as an intellectually valid stance without implications for religious belief, as many proponents claim it is. I think I can see how and why they make that claim, although I am still confident they are wrong. On the other hand, I've been surprised to discover that there are several definitions of evolution that I do believe represent the truth, and are compatible with my belief in divine creation.

There will always be those on both sides for whom it is enough to shout and harp, hoot and holler, mock and preach. As long as there are some behaving like that on one side, we can expect some to be like that on the other side. I think it is a foolish and inconsiderate waste of energy to be like that, and almost as foolish to try to reason with those who are. This book is for those on both sides who are willing to step back, look at the big picture, and see how our views took shape within their worldview frameworks over time, and how those frameworks continue to influence our evaluations of facts and positions of faith.

Step Two: What game are we playing? What field are we on?

To have a solid foundation for understanding each other and our differences of opinion, we need to dig down to the very roots of how we get our opinions in the first place. If you want to dig deep in this area, I suggest that you study philosophy, especially epistemology, the study of how we can know things. You may be very surprised by what philosophers have said about how difficult it is to know things. Some have gone so far as to propose we cannot be sure we know anything.

For the purposes of this book, I will look at three major ways in which we come to think we know things. These ways are direct experience, second-hand information, and logic. For a short and snazzy list, I will refer to them as Reception, Reports, and Reason. I will also use some computer terms that may help to show what I'm talking about.

When we are born, all that we learn comes from direct experience – our Reception through five main senses (if we're born with the usual complete set) and less well-known ones such as the ability to sense the tension of our muscles. We cannot yet understand what other people tell us and we have little ability to reason in depth. The processing of the signals from our sense organs is still developing. This Reception is comparable to the input a computer gets from microphones, scanners and cameras, but our brains do far more with it than computers can. As experts in the field of Artificial Intelligence have discovered, the programming needed to make sense of such input and operate in real-world situations is far more complex than that needed for things we consider "brainy" tasks, such as playing chess or answering trivia questions.

As our basic, subconscious logic develops and we start to make sense of the jumbled flood of signals flowing into our brains from our eyes, ears, etc., we gradually develop awareness that there are other people communicating still more information to us, information which we have not learned, and perhaps should not or cannot obtain directly. We learn early on that some of this information is unreliable: "Mmmmm, strained spinach! Yummy!" On the other hand, we also quickly learn that some of it is not only reliable, but very important and better than direct experience: "No, no! That's HOT!" Eventually, we learn to get still more information from Reports, as we read books, watch TV, and search the Internet. We may find a lot of information on things we can never experience directly. This, too, is sometimes unreliable: "Billy Bob's Snake Oil cures cancer and anything else that ails you!" It is also sometimes reliable and better trusted than personally tested: "Smoking is addictive and may cause cancer."

Finally, we become aware of the logic that has been developing all along, and begin to Reason in purposeful ways. A common example is the basic IF...THEN logic also used in most (if not all) artificial intelligence programs. We may reason, "IF I eat these cookies before supper, THEN I will get in trouble." We can program a computer to respond when a user enters "Goodbye":
IF INPUT$ = "Goodbye" THEN PRINT "Sorry to see you go!"
Of course, humans can reason and function in the real world in ways far beyond what any computer can do, even though a computer may be able to give the correct answers to far more trivia questions than any human.

It may be that however much we develop our understanding and use of formal logic, it is always connected to and influenced by the way our brains developed through experiences before we could talk or understand others, and by what we were taught by others. So, while our ability to Reason is the basis for discussing how we have arrived at our positions on various issues, our reasoning, is shaped by the information we have Received directly and gathered from Reports. Each of these ways of acquiring information has its advantages and weaknesses.

Reception through our senses is direct and immediate information, but requires some interpretation. Any Report that comes to us is depends on the reliability of the reporter. Our Reason works with whatever data is provided, plus memories of past experiences. New experiences are judged to be amazing, suspicious, weird, scary, etc., based on previous experience. That is why children's lives are filled with wonder, and relatively rare things like rainbows and helium balloons can still fascinate and cheer us.

But our senses are also well-known to be subject to error. We know that our perceptions may not accurately match reality. We see things like mirages and optical illusions. We judge how bright or dark an object is by the brightness of its surroundings. A similar alteration of judgment occurs when we have been exposed to heat or cold and then attempt to guess the temperature of something else. And our direct Reception of things is limited to the small part of the world within our reach or sight, and the limited range of the electromagnetic spectrum that we perceive as light. Some things are too dangerous to look at, touch, smell, listen to, or taste. Instruments like telescopes and microscopes extend our ability to directly sense some things, but they have some limitations, too.

Reports from other people and from devices we make to sense things for us give us information that we cannot gather directly. We can learn about other parts of the world. Thermometers report the temperature of things that would burn us. Telescopes with cameras show us the Sun without blinding us. Space probes report conditions far away from Earth. Perhaps most importantly, reports can tell us about the past, hundreds and thousands of years ago, where neither living human nor machine can go.

The disadvantage of reports is that we have to trust someone or something else to do the sensing for us, and report accurately. There are ways to check the accuracy of some reports, but not all of them. If we are too trusting, we may take in false information, but being too skeptical may lead us to reject accurate information.

Reason helps us, on a basic level, to make sense of our experiences and to consider and understand information reported to us. On another level, it gives us the ability to make generalizations and think about abstract concepts, mathematics and formal logic. Reason is the only faculty of gaining knowledge that produces new information apart from an outside source – once information from outside is gathered for Reason to work on. Using the power of Reason, we can extend the information we receive, and predict (with varying degrees of accuracy) events that haven't happened yet – tomorrow's weather, the path of a space probe, etc. Newton's work on momentum and gravity is a classic example of the power of reason applied to repeatable observations.

Reason also has its disadvantages. It has to work with the material fed to it through our senses and the reports of other people and our artificial sensors. Whatever errors may be in the other two sources may cause our reasoning to be flawed. In computer data terminology, "garbage in, garbage out." The power of Reason to manipulate generalizations and abstract concepts also gives it the freedom to produce logical, self-consistent results starting from observed facts – but the results may have nothing to do with the way the real world works. Some people use their ability to reason to deceive other people. People use their reasoning when they are "in denial." We commonly rationalize our own desires and bad behaviors, and justify wrong decisions. As the Bible says (Proverbs 26:16) "The sluggard is wiser in his own conceit than seven men that can render a reason." No statement about the world around us derived by logical reasoning can be considered a fact until it has been tested by observations, by other people, by comparison with other possibilities, and by the results of accepting it as true.

So we see that our knowledge and understanding of the world relies on all three channels of information working together and providing checks on each other. Everything comes to us through our senses, but Reception without Reports and Reason would leave us at the level of lower life forms. Reports give us information we couldn't get by any other means, but they need to be considered by Reason and consistent with other reports and our own Reception. Reason works with the information from Reception and Reports, and in turn, its conclusions need to be tested and connected to the real world by them.

An awareness of the uses, weaknesses, and interdependence of these different ways of learning about and understanding our world is the first thing we need when considering any subject with proper caution and humility. Different attitudes towards different sources of information are behind all the different views of creation and evolution. Does the Bible contain a clear and reliable Report of Creation? Is acceptance of the Bible's Report the only Reason to believe in Creation? Does what we Receive directly support evolutionism or creationism? How much can we Reasonably say we *know* about the past, especially in cases without any Reports of those who were there to Receive (from different viewpoints) what was going on? There are all sorts of positions, from those who accept the Biblical Report of Creation and aren't interested in considering any other sources, to those who are convinced that all we Receive about the world and our Reasoning together prove that Evolution (everything evolved from pure energy by purely natural processes) is a fact so obvious that anyone would have to be crazy to deny it.

It is beyond the scope of this book to do justice to all the issues raised in these considerations. I do hope to make readers aware that our choices between Creation or Evolution – or combinations of the two – depend on personal experiences, training and preferences of which we're often not aware. Often we find ourselves shouting at each other and wondering why the other guys aren't getting our message, because we don't realize that we're arguing from different mental ballparks. This awareness should spare us some trouble.

I believe that at the heart of the debate is a preference for one of two philosophical assumptions which are so basic that they cannot be decided upon by Reception,

Reports, and Reason (technically, they are axioms). The first was that what we observe is merely a subset of a greater reality. The second is that the natural reality we observe contains all the information we need to know all the important facts about it, including its entire history. The two are not perfectly opposed to each other, but that merely adds to the confusion and continuation of the debate.

Step Three: Draw the lines on the playing field clearly.
(Oops! too late!)

Before moving on to chapter two and the history leading up to the debate, let's look at a specific example of how something as apparently clear and simple as a definition can lead to confusion and conflict that we can never seem to clear up. Defining terms is arguably the most important issue to work out before beginning to debate. If we can't agree on definitions, or if either side switches between different definitions without giving notice and explanation, we can't hope to understand each other, let alone debate constructively.

To start with, the key words "creation" and "evolution" each have several different definitions, and this alone contributes much misunderstanding to the debate. Fortunately, there are some definitions that everyone agrees on, although they usually aren't relevant to the debate. For example, nobody has a problem with the idea that humans create works of art and many other things, and nobody argues there is no evolution in the general sense of "change."

When it comes to the disagreements, however, there are all sorts of positions. There are even disagreements over what the different positions are. Some groups claim to hold to a position and other people say they don't belong, while some people slap labels on people who deny they hold those positions. We can draw lines to show what we think the different positions are, but in practice the field is already a confused blur of dust as people draw different lines and others scuff out those lines and draw their own. I hope you will put up with me as I go to some length to describe the situation as I see it.

The traditional and most distinctive concept of "creation" in the context relevant to this book is the "literal" (plain, straightforward) Biblical young-Earth creation (YEC) position. This says that we can take the descriptions and references in the Bible to divine Creation of the heavens and the Earth at face value, that is, that God directly created everything in six days, about six thousand years ago. At the other end of the spectrum of ideas is the atheistic, reductionist evolutionary view of the universe, which says that there is nothing supernatural and (after the universe suddenly appeared) everything, including all forms of life, gradually took shape by natural, unguided interactions of matter and energy. The playing field is covered with a full range of positions between these two. Adding to the confusion, it often happens that teams close together at the same end of the field fight vigorously against each other.

Starting from the creation end, the YEC position includes options for the universe outside the Solar System, and even for everything except for the extant kinds of life on Earth, to be very old. Some allow for quite a few more than 6,000 years after the creation of Adam and Eve. Then there are various old-Earth creation (OEC) beliefs which also claim to accept the literal Biblical position, while accommodating the standard dates generated by uniformitarian assumptions. It can be hard to tell if a position is young-Earth or old-Earth. What they all have in common is the belief that God actively created (or re-created) the Earth and the first forms of life.

Spreading across the middle of the field are a lot of other groups with ideas about origins. Progressive creationism accepts most, if not all, of the standard dating and sequence of appearance of living things, but holds that God intervened at key points. Some progressive creationists also argue that their position is consistent with a literal reading of Genesis.

The Intelligent Design (ID) movement is a "big tent" group that includes everyone from YECs to theistic evolutionists, as their only point is that aspects of nature, especially living things, exhibit a type of complexity that doesn't happen by chance. I think it's very revealing that strong evolutionists insist on lumping the whole group with Creationism.

Theistic evolutionists (TEs) cover the evolutionary end of the field right up to the full atheistic evolutionary position. Some may hold that God has constantly and actively guided evolution, but only by undetectable influences on Nature. Others merely add a footnote or two to the atheistic position, holding that God set things off "just right" at the start, and intervened near the end (and then only on a spiritual level) by giving humans souls. TEs and atheistic evolutionists set themselves apart from each other, but in the debate about the story of universe, they gladly team up against YECs.

The typical official position of major scientific organizations is one of neutral, naturalistic methodology ("science"), but in effect it is a naturalistic philosophy that is effectively the same as Atheistic Evolutionism. The acceptance of the modern view by many churches, and the inability (or refusal) to see the philosophical bias in this area of science partially explains why the debate is so often portrayed as a matter of fundamentalist religion versus science in general. Mostly, however, it suits the goals of leading atheists. Late in its history, the term "science" often has been used with more propaganda value than definite meaning. Any claim that is said to be scientific automatically gains prestige.

There are sometimes splits and nasty fights within the extreme atheistic evolutionary end of the field. To the rest of us, these seem to be a lot of fury over small differences. The major scientific establishments handle such disputes by: 1) Restricting them to scientific journals and keeping them out of the popular media as much as possible, 2) Strongly attacking anything which might support or give comfort to any position at all closer to the other end of the field, 3) Eventually squelching the new idea entirely if possible, 4) Toning down or otherwise changing the new idea so that it is clearly consistent with naturalistic philosophy.

Some significant debates have been over theories and discoveries that went against the simplistic extremism of the geological theory (uniformitarianism) which had opened the door for evolutionism. Arguments and evidence presented in support of plate tectonics, and J. Harlen Bretz's proposal that a large flood or floods created the channeled scablands of Washington, challenged supposedly well-established, fundamental "facts" of geology. The Big Bang theory also had some opposition (which hasn't completely died down), in the spirit of uniformitarianism and the general naturalistic philosophy behind the current practice of science.

There was a strong negative reaction to the system of classification known as cladistics, because the evolutionists who came up with it said that it wasn't necessary to consider how living things might have evolved in order to organize and classify them. After swearing their allegiance to evolution and pointing out that their classification charts could be understood as showing evolutionary relationships, they were welcomed back into the fold.

Other debates are about whether or not there is some sort of progress or direction in evolution, and to what degree evolution is controlled by competitive natural selection rather than cooperative interactions such as symbiosis.

Perhaps the most famous skirmish was when a group broke away from the traditional gradualist neo-Darwinian camp and advocated Punctuated Equilibria, which said that the major steps of evolution came in short, widely-spaced bursts. Not only was it less uniformitarian than traditional neo-Darwinism, it argued (as do creationists) that the fossil record does not show a gradual evolution and that major steps are missing. They said this was because evolution sometimes happened so fast that there wasn't much chance of the transitional animals getting fossilized. Such extreme changes in evolutionary rates and the admission of gaps in the fossil record were too much for many evolutionists. The debate got rather nasty and spilled out of the inner circles and into the attention of the general public. However, after it was noticed that creationists were pointing to the arguments with delight, the "Punk Eek" team watered down their arguments, and now the debate is little more than a hair-splitting squabble over a technical issue.

So there is a wide range of positions hidden under the labels "creationist" and "evolutionist," and in some ways they overlap a lot. All this is only a glimpse at how complicated the debate is. It is not a simple battle of Religion versus Science. Getting lost in all the heat and smoke are significant facts that could cool things off. One is that even YECs can accept several useful definitions of "evolution." Most informed YECs accept that there are shifts in the gene pools of populations, and that populations can split up into different species. Most of us also accept natural selection. YECs also agree that God could have created everything using nothing but gradual, natural, evolutionary processes – we simply believe there is good reason, Biblically and scientifically, to believe He started by supernaturally creating the heavens and the Earth pretty much fully formed. Nobody believes that the Bible is a science text, but YECs do believe it is historically accurate from the very beginning.

The most important point that has been lost in the confusion is that all the debating is based on one key point, which is neither scientific nor religious, but affects both religion and science. That's why an overview of the philosophy and history of the debate is needed. Creationists believe that the original methods of science (establishing facts about the workings of Nature by repeated observations and experimentation) provide evidence that strongly indicates that life could not have formed by natural forces alone, and that the potential for natural biological change (evolution) is limited and could never produce all life from microbes. This form of science remains compatible with the traditional religious view that God could and did actively create and intervene in the universe. It also leaves evolutionism as a faith without scientific demonstration.

Although anti-creationists have emphasized the religious opposition to Galileo's arguments for the current view of the Solar System, and promoted a myth about church leaders opposing Columbus because they believed in a flat Earth, the real trouble began later, with a philosophy that infiltrated both religion and science. The traditional Christian view is that God the Creator is in no way bound by the laws of Nature that He created. A newer view, influenced by excitement over the successes of science and renewed admiration for ancient Greek philosophy, concluded that once God created the laws of nature, any supernatural intervention would be a violation which He would not commit. This gave rise to the religious position of Deism. It also led to a parallel change in science. If natural laws could not be interrupted even by God, then science could reveal more than the way natural things normally worked, it could puzzle out the entire history since the Creation. This was a possibility some people found too exciting and satisfying to resist. As Christian denominations drifted toward accepting Deistic ideas, they began to reject the Bible's Report of history and accept human Reason (in the robes of the expanded view of science) as the dominant source of information about the past.

So that is the key issue: whether or not this philosophical expansion of science is valid, and to what extent its results reflect the truth. The scientific Establishment argues that science can explain everything in terms of natural processes, and any allowance of supernatural intervention is not science. It generally claims that this view has nothing to do with religion or philosophy, but no intellectual pursuit is philosophy-free, and evolutionists from Darwin on have appealed to religious arguments, and some have said that Evolution makes belief in God unnecessary or gives validation to atheism.

For those seeking an intermediate position, it comes down to how far they accept the power of naturalistic philosophy to enable science to reveal the past, how much they trust the appearance of independent reinforcement of secondary or circumstantial data analysis. After that, it's a matter of choosing some way to accommodate their religious beliefs to that position. For that matter, debates among evolutionists sometimes depend on variations in confidence in the conclusions of science that depend heavily on naturalistic philosophy and can't be confirmed by repeated observations and experimentation.

Unfortunately, the importance of this fundamental issue is lost in the complexity of technical details, debates over specific issues, and the confusion created by words having different meanings. Perhaps not surprisingly in a field where merely trying to identify the different teams can be confusing, there are a number of important terms with different definitions that are constantly being argued over.

For example, creationists may say they accept "micro-evolution" but not "macro-evolution." What they mean is that they accept variations within "kinds," but not the evolution of new kinds of organisms. To an evolutionist, however, micro-evolution may involve only very slight variations within a species, while macro-evolution is the formation of a new species as a result of micro-evolutionary steps over time. An evolutionist may assume that speciation demonstrates the ability of living things to evolve from a simple common ancestor into all other forms, given enough time. So two people can be arguing about these terms but have very different ideas of what they're arguing about. There are both evolutionists and creationists who argue that the terms should be avoided. Evolutionists don't recognize the term "kind" because they don't use it, don't believe in the concept, and creationists are still working on defining kinds, but there is some controversy over the term "species," as well. Other classifications are acknowledged to be somewhat arbitrary but useful artificial groupings, and scientists can be "lumpers" (combining very different things in the same classification) or "splitters" (dividing things with slight differences into different classifications).

A similar and related problem of definitions is the question of "intermediate" or "transitional" fossils. Evolution of new forms of life is said to proceed too slowly to observe directly, even in creatures that reproduce very rapidly, such as bacteria. Fossils are supposed to provide a sort of indirect observation of evolution. As Darwin noted, if his belief in evolution were correct, there would be a smooth series of very slightly different forms of life from a few simple forms right on up dinosaurs, whales, humans and everything else that ever lived.

The fossil record shows many different forms, but the vast majority of them are clearly not somewhere along a line of evolution to extant things – they are too different from, or too much like them. Thus, fossils that can be called intermediate in form between more primitive life and modern life are eagerly sought, and finds are loudly proclaimed. However, in most, if not all, of these cases, they are still too different, too specialized, or dated too recent to be an actual ancestor of a living kind. So almost always, "intermediate" is a general term and "transitional" refers to a fossil which is something like a hypothetical transitional life form. Not only is "transitional" shorter than a phrase describing the position it is believed to occupy, to a believer in evolutionism, something close enough is good enough evidence that the transition actually occurred.

One last major area of confusion is in the phrases "scientific fact" and "scientific theory." You will find books from various positions which claim different things, from "evolution is a fact" to "science isn't about proof" – perhaps in the same book. This is complicated by the fact that "evolution" is sometimes defined in a way even YECs could agree with, and sometimes in a way hardly compatible with any theistic religion. Often, debate rages over the way creationists point out that evolution is a theory, not a fact. Most evolutionists will agree to this, but point out that scientific theories can be so well supported that they are generally regarded as fact, such as the "theory" that the Earth is round (spherical), the "theory" that it travels around the Sun, and the "theory" that gravitational effects are described by Newton's formula (with some adjustments thanks to Einstein).

However, we seem to have two different kinds of theories here, along with two different definitions. Theories like Newton's theory of gravity and Einstein's theories of relativity have mathematically precise consequences and make similarly reliable predictions. They can be and have been directly observed and demonstrated to be true by many different experiments. When the theory of evolution is clearly and precisely defined in scientific terms, it deals with the variations between generations which have been observed. However, in general terms, the theory of evolution is understood to mean the belief that all living things descended from one-celled ancestors. Of course, this vast story is beyond observation. Evolutionists will argue it can be indirectly observed by circumstantial evidence, or compare its study to forensic science, but creationists will point out that forensic science involves matching evidence to possibilities that have been carefully observed and tested, not describing events of a sort that have never been seen.

Of course, I could be mistaken on some points, as I pointed out at the start. I am at least aware that evolutionists have their reasons for overlooking or dismissing any shortcomings I have pointed out. I've had correspondences with evolutionists in which we ended up getting frustrated with each other because our viewpoints prevented us from seeing the same data as meaning the same thing. Recognizing the philosophical gap in what should have been common ground helped us remain civil. In the process, I've seen glimpses indicating that if we focus on what we can truly agree on, and cautiously proceed from there, we can at least improve our separate theories and arguments, although they remain opposed.

Seriously considering ideas that you think are ridiculous or impossible is a good way to explore the limits of what is and is not possible, and why. Considering challenges to what you believe, examining the reasons you can give to fend off doubts, and walking in someone else's shoes can give you a better understanding of where they're coming from. It can also give you a greater appreciation for your own beliefs. There may be limits to the number of ideas you have time to examine, but before you decide which to ignore, be sure your decision is based on truly established facts, not "facts" that are ultimately beliefs dependent on faith and philosophy, even if mounds of data are gathered in support of them.

The idea that living things (and everything else) were created by God (or gods) came down to modern times in the form of a religious dogma, long before there was any organized scientific endeavor. But where did religion and belief in divine creation come from? There seems to be a popular set of ideas about this which are sometimes stated as factual or highly probable, but what do we truly know? This is an important part of the background of the debate, and it is one of the things we'll look at in Chapter Two.

Chapter 2: Where did we get our ideas from?

Section One: Looking backward through the mists of time.

Understanding the philosophy of knowledge and the debate is the best way to start studying a controversial subject, it is also important to know the historical background. The problem with this particular subject is that the two positions come with two different views of the history of the Earth! Ideas about what we can know about the past are taken as basic axioms vital to the existence of science, and beliefs based on the interpretation of scientific data are considered historical facts. This view leads to strong emotional responses to the young-Earth view, even from other theistic positions. Naturally, it's hard to have any respect for a view that you believe was dreamed up by ignorant humans only recently civilized. Even those who believe the Bible is inspired by God show great disdain for YECs, largely because they believe that God made the world so that the naturalistic assumptions of science provide clear and trustworthy knowledge of the past. Once they have accepted this view, the tendency is to eventually see the Bible's account of Creation as merely an instructional mythology or vague poetry that has little, if any, correspondence to the actual history of the Earth before people began writing and saving historical records.

In our schools, science journals, and even television shows and movies, that humans evolved gradually from ape-like creatures over millions of years is taken as a fact, beyond question. It seems that many people believe that religion also gradually evolved, that the Greeks started science, that the rise of Christianity caused the Dark Ages (until the re-discovery of the Greek texts re-kindled religion-free scientific thought), that early scientific men like Galileo, Christopher Columbus, and Francis Bacon were champions of Rationalism who fought against the ignorance of Religion, and finally Darwin discovered natural selection, leaving no earthly reason to believe in God. I may be exaggerating and oversimplifying a little, but I've seen letters to editors and comments on web pages saying similar things.

How can anyone believe God created the Earth a few thousand years ago, when "everybody knows" (because "science says") that our species has been around for about 300 thousand years and the Earth is billions of years old? How can anyone else regard such a belief as any more than stubbornly sticking to an outdated religious belief? It's not so hard when you consider what we've said about knowing things, and apply that to what we think we know about history and prehistoric time.

One problem with history is that we can't go back and live it again; it is beyond our personal Reception. The vast majority of what we know about history relies on the Reports of people who lived through those times, although most are known only from copies made long afterward. This must represent only a small fraction of all that went on. People who were famous in recent history are already becoming forgotten despite all the records we have and the ease of finding them on the internet. Do you know about Jenny Lind, Enrico Caruso, Mario Lanza and Nelson Eddy? How about Sergeant York, Private Roger Young, Dr. Albert Schweitzer, or Luther Burbank? If you do, ask some young people if they know them.

As with all reports, the information we do have can be biased. As the old saying goes, "History is written by the winners." But there are odd alternative viewpoints, too. Consider the differences of opinion on the situation in Iraq in 2012, the conspiracy theories about 9/11, and about the assassination of JFK. Let's step back through history and see how opinions have changed, and how events and even kingdoms become shrouded in the mists of time and uncertainty.

For example, before 1938 scientists believed that all coelacanth fish had been extinct for millions of years. Then a scientist found out they were being caught off the southeast coast of Africa. We now know there is more than one species and they live in different places. In the late 1800s, it seemed to many people that science had settled all the major questions and soon there would be nothing left to do but fill in the details. There were just a few strange things that the most advanced researchers were puzzling over. Little did they know that in the coming years there would be revolutions in fields such as physics (general and special relativity, quantum dynamics), astronomy (the discovery of galaxies and universal expansion), and biology (genetics and DNA). At the end of 1859 Darwin published his book that started the modern belief in evolutionism.

The publication of the King James Authorized version of the Bible, bishop Ussher's estimation of the age of the Earth as about 6,000 years old, and the Roman Catholic church's suppression of Galileo's work showing the Earth going around the Sun along with the other planets, are associated by many with a dark time when religious forces fought against science, but at the same time, Kepler was producing works on heliocentricity without opposition in the Protestant region where he lived. Sir Francis Bacon published works describing the scientific method (as the proper procedure for natural philosophy) and calling it secondary to the word of God and "given to religion as her most faithful handmaid." A number of other early advances in science were underway also, such as Torricelli's invention of the barometer.

When I was a child, we learned the little rhyme "In 1492 Columbus sailed the ocean blue," and a commonly-told story (as seen in the Bugs Bunny cartoon, "Hare We Go") said he first had to argue with ignorant religious people that the Earth wasn't flat. We now know the real problem was that he had an unrealistic notion about the size of the Earth, and educated people knew the Earth was a larger globe, so it would be much farther to China than Columbus thought. Martin Luther's 95 theses, posted in 1517, provide a milestone for the Reformation, but if we look about another hundred years further back, we see there were already some early reformers, such as Wycliffe and Huss. The Renaissance was also starting about that time.

Five hundred years (in round, rough figures) before Columbus, one thousand years ago, brings us to Leif Erickson and his once-forgotten voyage to what he called Vinland. Looking over the period stretching back to 1500 years we already see characters we know more by legend than history, such as Robin Hood, King Arthur, and Beowulf. This stretch of time also covers the fall of Rome, and Muhammad and the rise of Islam.

We have to jump back another 500 years to trace back over the time when Christianity first became the official religion of the crumbling Empire, the earlier persecution of the Christians, and finally the birth of Christ. As significantly as Christianity has affected the history of the world (Europe and the Americas in particular) many theologians doubt that anything written about Jesus of Nazareth is accurate, and some people doubt he ever existed! (Practically all serious historians, however, acknowledge he was a real person.)

A device called the Antikythera Mechanism, dated to shortly before this time, may suggest how spotty our knowledge of the past is. Nothing like it from its time was known, until it was discovered about 100 years ago on a sunken ship. Only recently was its purpose and significance figured out. It appears to be a mechanical computer for calculating the positions of the Sun and Moon, and possibly the planets that were known then. No mechanism of similar complexity would be created for over a thousand years, as far as we know now. It seems to be one of the last advances of Greek science and philosophy, which had long before passed its climax. In fact, it was almost 170 years before Christ that the Roman Empire began its expansion, replacing the Greeks as rulers of the Mediterranean area.

The rise and flowering of the Grecian empire and its greatest advances in philosophical math and science seems to have pretty much run its course by 2,200 years ago, having begun about 2,600 years ago (with Pythagoras). It was during this period that the earliest history was written by Herodotus, who is thus known as "the Father of History," but he has also been called "the Father of Lies" because of the disputed, legendary and mythical elements in his work. His history (and many other non-fiction documents long after) included supernatural events. We now have only parts of his work, from copies made long after his time. All history before this is known only by assembling smaller bits and pieces; for example, monuments concerning one event or the reign of one ruler, from one point of view. Most, if not all, include references to the supernatural. The war between the Greeks and the Persians may be the oldest event that we know of from two different non-religious sources with different viewpoints.

Also in the period between 2,500 to 2,600 years ago we find the earliest known circumnavigation of Africa, the founding of Jainism, Siddhartha (Gautama Buddha), Kung Fu-tse (Confucius), Daniel, and before those, the destruction of Jerusalem by the Babylonian Empire.

About 2,700 years ago lived a king known mostly as a character in a fantastic myth: King Midas. Apparently he was a real king, but the stories about him made up by people in rival countries are his greatest legacy. Another 300 years ago brings us to three thousand years ago, and the time of King David. David, and his entire dynasty, also have been considered mere legend, but archaeological discoveries have supported the existence of David's kingdom, and of a famous Philistine named Goliath. Not surprisingly, the significance of such finds have been disputed.

Another 300 years back sees the climax of the Phoenician Empire and the reigns of Pharaohs Ramses II and III. Also during this time would be Joshua and the fall of Jericho, and the Trojan War. Again, both the fall of Jericho and the battle of Troy have been considered pure myth by many in the past, but there is evidence for the existence and conquest and both cities.

Events before this are known only from fragments, brief inscriptions and hieroglyphics. Anything before this period is arguably prehistoric, with nothing more than the equivalent of a scattered trail of crumbs to provide Reports for our knowledge about the past. While historians, archaeologists, and others may dispute this claim, it seems to me that it makes sense to emphasize the uncertainty of any statements about such ancient matters.

At any rate, it seems that from this point on, there may be some doubt about the exact date or some other aspect of any event. Some scholars have doubted there is significant historical truth to the account of Moses and the Exodus of the Israelites from Egypt, let alone the miraculous events associated with it. There was a theory that Moses couldn't have written the Pentateuch because he couldn't have known how to write, but that was proposed before discoveries of earlier writing.

So it seems significant that history, as a solid line of recorded witness accounts, fades away as we approach the time of Noah's Flood – and many cultures have similar accounts of an ancient global catastrophe. Over the past 4,000 years, cities and whole civilizations have been lost. We know of some that were re-discovered. Scholars have suggested there are threads of truth in the story of Atlantis, claiming they lead back to some real civilization that fell and was lost to history. One candidate is the Minoan culture, which ended about the time a volcanic explosion totally destroyed most of the island of Santorini. One or two investigators have proposed that side effects of this explosion also affected Egypt and inspired the story of the ten plagues of Exodus. Among many wild theories about Atlantis, another serious proposal is based on evidence found at the southwest end of Spain.

A definitely real lost civilization was that of the Hittites. Its absence from the body of known history was for many years pointed to by those who denied the historical value of the Bible. The discoveries of other references to, and archaeological sites of, a people matching the description of the Hittites gives vindication to the biblical account of the time, the time of Abraham and early great cities such as Ur of the Chaldeans. There were also great civilizations that fell and were lost for centuries in India, Africa, and Southeast Asia.

So, just about where secular history parts ways with the part of the Bible that young-Earth creationism is based on, there is no other reliable written history! Dates assigned to things earlier are derived from scattered bits of writing that can't be accurately dated, and dating methods that assume the steady, uninterrupted working of natural forces, disregarding the accounts of major supernatural events in many ancient writings.

We "moderns" may find a great deal of comfort in thinking that the world has always been as mundane as it is now, and that our scientific methods can reveal almost anything to us, even showing us truths about the distant past, as if we could travel back in time. But consider the points raised in the first chapter: How much can we truly know about such things? Can we totally disregard Reports from the past, or accept parts while throwing out everything else?

Then, too, while there are good arguments for connecting the events in Genesis to certain dates, those dates are not explicit in the Bible. Perhaps there is some leeway there as well, although it must be very small compared to some of the conventional dates of the supposed prehistory of mankind. However, various things have been dated with modern techniques, only to be given much different dates by later tests or different techniques.

Another interesting point is that all of the first great civilizations, in Mesopotamia, China, India, and Egypt, are all dated to a period less than 6,000 years ago. As we look at the story of prehistoric humans, we find that settled cities have been dated 9,000 years old, carved stone monuments thousands of years older, and anatomically modern human skeletons have been dated over 100,000 years old. Why did it take so long for such civilizations to form, but then advance so rapidly in those areas, and why not in Europe or Africa? Why are the oldest settlements in the western hemisphere also dated to the time of ancient Egypt?

Consider all that has been accomplished in the last fifty or one hundred years, let alone in the 500 years since the scientific revolution began. Think how quickly popular fashions change. Consider the powerful countries and empires known to have risen and fallen over the centuries before modern technology. Think of how quickly humans can multiply and spread when conditions are right. Think about how uncertain we are about events in the historical period.

So, while it may seem hard to believe that the Earth is only a few thousand years old, there is only a few thousand years' worth of written human records. Now we can look at the standard human prehistory and wonder why so little happened for tens and hundreds of thousands of years. Let's give it a try.

Chapter Two, Section Two: The Story of Human Prehistory (or What Took So Long?)

Where you choose to begin the story of humanity depends on your beliefs about origins in general. Evolutionists, and some creationists, too, might begin over 13 billion years ago with the Big Bang. Or the first life, over 3 billion years ago. How about the first ape? Still dated over ten million years ago.

But for now let's focus on what we might call the first human, perhaps the first beings to have their overall anatomy closer to ours than to that of living apes – big brains, able to walk upright for long distances easily. How long ago are they supposed to have lived? What are the supposed stages of their development up to the first great civilizations?

Given that focus, we'll skim over the story of how some common ancestor of all the great apes (supposedly including humans) some 20 million years ago had many different ape-like descendants, and how the line leading to humans split off from the rest somewhere between eight and five million years ago. We'll skip over the australopithecine apes, said to have walked upright and to have used stones as tools, but still more ape than human, as the "pithecine" indicates. What is the date given to the first fossil placed in our genus, *Homo* (Latin for "Man")? The dating and the classification of the fossils may be a bit controversial, but we might start the story about two and a half million years ago, with *Homo habilis*.

It is estimated that this species had brains of 600-700 cubic centimeters, about half the size of a small brain in normal modern humans, but it is unclear how relevant brain size is to intelligence. At any rate, it's quite a bit larger than that of the creatures we have fossils for that are considered closely related to its (and our) ancestors. But this start is rather shaky and hazy. It seems that the first sign of human activity that is distinct from that of apes are some small, very simple stone tools, mere pebbles with a few chips, associated with animal bones showing cut marks, possibly due to butchering. These simple stone tools, known as the Oldowan toolkit, are dated to about 2.6 million years ago.

Now here at the start we have the first example of "What took so long?" Supposedly someone or something had developed the brains, manual dexterity, and coordination to figure out that stones chipped just right could be useful, and then make and use them. Supposedly this happened over two and a half million years ago.

So, how much time supposedly passed for this innovative individual to figure out that a different pattern of chipping might improve the cutting edge or make a rock useful for some other purpose? Well, you can forget about the first rock-chipper figuring that out. Nor the generation that followed. Nor the next species. In fact, the first big advancement in stone tool-making, called Acheulean tools, is given a date about a million years more recent. And the most recent of them are dated about a million years younger still!

These millions of years are far beyond our experience and historical records. Looking back as much as four thousand years is a strain that takes us to times shrouded in myths. But let's round up and take five thousand years as a ballpark figure, and call it one historical period or "one history" for short. Now we use this as a measuring stick for comparison. This shows us that the first stone tools are dated over 500 histories ago! About 19 histories are lost in rounding off. The million or so years before the Acheulean style was invented covers 200 histories. Isn't that an incredibly long time for something like that to happen?

There are some indications that these early humans, *Homo erectus,* were quite intelligent and resourceful. Not only are their fossils found over a wide range, including north of the 40th parallel, they are also found on islands off the coast of Asia. The water level at one time in the past was much lower than it is now, but apparently they still would have had to build boats or sturdy rafts and go to a lot of effort to get to the islands. There are also some artifacts that indicate* *Homo erectus* could think in symbolic terms – stone "tools" too large to use as tools, beads, and patterns scratched on stones. (*According to the dates assigned to them, there were no *Homo sapiens* around at the time.)

Another significant step toward modern civilization – traces of shelters – don't show up until about 500 thousand years ago. That's another 200 histories without significant advancement. We might say that things start getting interesting at this point, although we've still got almost 99 histories to go before the first major civilizations. According to estimates, Neanderthals are thought to have shown up about this time. Some wooden spears found in Germany are dated to 400 thousand years ago (we've now jumped 20 histories as if they never existed) and there are traces of what appear to be fireplaces, too, but it's not at all clear which variety of humans created them.

Skeletons assigned to our species, with our bubble-shaped skulls free of the Neanderthal brows and such, are said to be almost 200 thousand years ago (another 40 histories, with 39 still to go). Oh wait, newer ones are dated about 300 thousand years old, still 20 histories farther on, and almost 60 histories before we get to actual history! In 2017, re-analysis of fossils thought to be Neanderthals that died 40,000 years ago indicated they were actually a lot like us, with a few "archaic" traits, and they were dated around 300,000 years old! (*Nature*, 2017; 546 (7657): 289 doi: 10.1038/nature22336, 8 June 2017) Given about the same date (320 thousand years ago, but what are 4 histories in dates like these?) are the first "Middle Stone Age" tools. Supposedly, pre-human ancestors had figured out how to make stone tools *two million* years earlier, and members of our genus improved on them about *one million* years after that, but otherwise, in all that time, never figured out how to make them better or make different styles and uses. Neanderthals had slightly larger (though differently-shaped) brains and interbred with our ancestors – we still have some Neanderthal genes. But the technology story remains: stone shelters, signs of using fire, some artifacts that seem to have symbolic meaning, and finally another improvement in stone tools. Not much to get excited about, considering how much more time passed.

This seems like a good place for a reminder that I'm using the accepted dates only for the sake of argument. More than that, however, I should point out that even those who accept the standard view should keep in mind that these are rough figures, subject to large changes, and many points may be controversial even among evolutionists. There's no telling what is being presented in science magazines as fact today that will be considered outdated and discarded tomorrow.

For example, some researchers found a site with many footprints in Mexico that included what looked like human footprints. The researchers used modern techniques to date the site as 40,000 years old. Other researchers had a problem with that, as they believed that humans didn't come to the New World until around 13,000 years ago. So, some of them used another dating method to check, and came up with a different figure – but not more recent, much older: more than a million years ago! This was so far from what they could believe, they suggested that what looked like human footprints in this case were probably something else. (see "Geochronology: Age of Mexican ash with alleged 'footprints'"by Paul R. Renne *et al.*, a "Brief Communications Arising," *Nature* 438, E7-E8, 1 December 2005). So, further study was done and indeed, it turned out that they weren't footprints at all! ("Techniques for verifying human footprints: Reappraisal of pre-Clovis footprints in Central Mexico" Quaternary Science Reviews 29(19-20):2571-2578, September 2010, Authors: Sarita Amy Morse, Matthew R. Bennett, Silvia Gonzalez, David Huddart DOI: 10.1016/j.quascirev.2010.03.012.)

Think that's just a fluke? For another example, it was once thought that modern humans and wild dogs or wolves might have begun moving toward civilization and domestication together, over 100 thousand years ago. It seemed at least probable that the relationship would have begun no later than the rise of modern humans to dominance over Neanderthals about forty thousand years ago. More recent studies suggest that most modern dogs are descended from wolves that were domesticated no more than fifteen thousand years ago. The vast majority of modern breeds go back no earlier than the 1800s.

Still not impressed? There are other examples. Some researchers were very excited when human artifacts in Australia were dated at over 100,000 years old. Another dating method later produced the ho-hum result of less than ten thousand years. *Homo erectus* was considered to have gone extinct about 200 thousand years ago, but when some water buffalo teeth were studied and dated, it indicated that a population of "Java Man" had lived more recently than 60 thousand years ago. So keep in mind that all of this is subject to change and new reports may have already altered some things.

Jumping ahead another 24 histories(!), there are some tantalizing bits of archaeology dated between 80,000 and 50,000 years old (a span of 6 histories) – better evidence for symbolic use of color and patterns, advanced spear tips of bone, hints of care for the very sick and old, the appearance of formal burial and other evidence that could indicate religious thought. Again, many of these are attributed to Neanderthals and some to moderns. Then there are tantalizing and more-or-less controversial finds suggesting that humans had sailed to Australia, and that some had made it to the Americas.

Around 40,000 years (8 histories) ago, there's some more improvement in stone blades, a lot more beads found at more sites, and a site indicating that some people (quite possibly Neanderthals) were managing to live in the Russian arctic. Modern humans supposedly began moving into Europe about this time. Between forty and thirty thousand years ago, cave painting began, and there is evidence that glue was made and used to join a stone knife to a wooden handle – both of these appear to be the work of Neanderthals. There are controversial claims that the Neanderthals made flutes. The dating of cave art has also been disputed, with some researchers claiming that at least one site was only fifteen thousand years old. Some cave paintings in Italy that were first dated about fifteen thousand years old were re-dated and declared to be around twenty-five thousand years old. Evidence for grinding and eating wild grains is dated as far back as twenty-three thousand years ago, but the oldest date suggested for farming is almost a whole history later, while other estimates suggest it took two or three histories before people began to settle down and grow plants for a living.

Fire-based technologies show a similar hint of stretched dating. Objects showing the use of fire to create relatively advanced materials such as glues and ceramics have been dated tens of thousands of years old, but fire-produced pots and vases do not become commonplace until a few thousand years ago.

About twenty thousand years ago, the height of the last ice age began and the oldest-known boomerang was made (in Europe, from a mammoth's tusk). If the dating is correct, around seventeen thousand years ago cave paintings in France were created that show about the same level of talent and artistic sense as cave paintings dated about thirty thousand years old or more.

So there are, here and there (but mostly in Europe and Israel), signs that early modern humans (and Neanderthals) were as clever, imaginative, and perhaps as spiritual as any of us, but living as nomads. Of course there are many more artifacts and traces than these examples might indicate, such as stone lamps that probably used fat or oily animal renderings for fuel. And humans built just like us had been around for about 300,000 years.

The last ice age went from its peak to its end in about one history, roughly from 16 to 11 thousand years ago (some say we are still truly coming out of it), farming began (according to some), and even the stodgiest of experts agree that humans moved into the Americas. North America at the time seems to have been swarming with all sorts of large mammals, which soon went extinct. There is some controversy over how big a part the new arrivals played in this. Considering the animals included large saber-toothed cats, brown bears, dire wolves, and wolverines that could have preyed on giant ground sloths, young mammoths, and enormous armadillos, we might wonder how the humans managed to compete and not get eaten. Apparently they did, which shows they were certainly clever and capable enough to form civilizations. The amazing site of Gobekli Tepe in Turkey is dated between 13 and 12 thousand years ago. It consists of many large stone pillars shaped into smooth T shapes with detailed carvings of animals on the sides.

Apparently the first solid evidence for domestication of animals (goats) is dated about ten thousand years old. Compared to the 149,000 years that it supposedly took modern humans to get to this point, there seems to be no time at all to go from goat herding to empires – but on the other hand, it's still a whole history of time before recorded history. Other bits and pieces of civilization pop up here and there about ten thousand years ago. For examples: The oldest actual remains of boats. Naturally mummified human remains. Traces of wine in China and the Middle East. Possibly the beginning of settlements and small cities in the Middle East, such as at Jericho, and in Jordan. An entire city, lost when it sank below the waters of the Gulf of Cambay, India (although in that case the dating is admitted to be uncertain). Small samples of marks which could be early forms of writing. Indications (or estimates) of domestication of more animals, and of farming spreading rapidly from the Middle East across Europe.

It's not until about seven thousand years ago, practically on the doorstep of recorded history, that we find the first actual fossils of what appear to be domesticated dogs. There is also the first collection of monumental stones that may have been related to astronomical observations. They are associated with apparently sacrificed cows at Nabta, Egypt. The first signs of civilization and the farming of a variety of crops in the New World are also dated to about this period. To me, at least, that seems too much of a coincidence to believe. It had been thousands of years, about one history, since humans had crossed a land bridge from Siberia to Alaska. So the standard story goes, at any rate. The environments and populations of the Americas and the Middle East would have been very different, along with the histories of the peoples, and yet the two developments of civilization are nearly synchronized.

To sum up: we're supposed to have had something like 193,000 years for physically modern humans to go from little more than stone tools to the first signs of towns and arranging monumental stones for purposes other than mere survival. The unobserved years are far too many for the traditional Biblical time-frame, but the known fossils, artifacts, and technical developments could easily be accounted for in a very short time. Some people have continued to use stone tools and live by hunting and gathering into modern times, so if our assumptions about dating ancient things are wrong (and we've seen some dramatic cases of changes in dating), then these artifacts could have been produced in a few hundred years before writing was invented, or even during historical times.

So now let's play another game and say all those artifacts and bones were produced mostly during the relatively brief period in dispute between the first signs of cities and the first great city-state civilizations. According to the standard dating this would be several thousand years (almost one history if we count a few early sites dated about nine thousand years old), while in the traditional biblical framework it was only several hundred years, right after the Flood. How much is supposed to have happened? Would it really need more time than recorded history? Or could it all happen in several hundred years?

Perhaps the case is somewhere in-between. There's certainly enough material (the civilizations of the Fertile Crescent, including the pyramids, etc.), but it's not stretched out evenly through that supposed time. Most of it, even in the standard dating system, happens at about the same time. Compared to almost two hundred thousand years of slow progress, it all pops up in the blink of an eye. Anybody who's done the calculations knows that a small group of motivated humans could populate the whole Earth in a few hundred years. If they carried the memories of a previous civilization they could educate and inspire their children and later generations to rebuild it, producing the pyramids of Egypt and the Americas and the ziggurats of Mesopotamia. A rapid rise of civilization not long before recorded history would also eliminate the question of why there are small samples of pictographs and symbols supposedly long before the first full and extensive examples of writing – they weren't actually that far apart. Our faith in our dating methods has been misplaced.

I'm not saying it's a simple, easy thing to reconcile the standard dates of early and "prehistoric" artifacts with the few thousand years since the Flood in the traditional 6,000 year time-frame. I'm merely pointing out there are some considerations which make it plausible that the truth is very close to the traditional view, and cast doubts on the currently-popular story. The relative lack of fossils and artifacts in the vast, imagined time from the first humans to the first large cities makes sense if our dating methods are producing systematic errors. The sudden appearance of farming and ranching, huge monuments that are still impressive feats of engineering, writing, formalized music with many kinds of instruments, the Middle East being the center of all this, and other facts, are all more consistent with the Biblical account than the old-Earth story.

As we saw in Section One, there are controversies over events of the twenty-first century, and the deeper past quickly becomes veiled by the mists of time. The idea that it is reasonable and proper to presume the world has always operated absolutely naturally is relatively recent, and arguably can't be tested. While some evolutionary ideas were proposed over two thousand years ago, variations of supernatural creation were the rule almost exclusively up until the 1700s. Historical records of all ages and cultures attribute major events to the aid or intervention of gods or other supernatural forces. There's little indication that people in early civilizations had any knowledge or perception of having many generations of ancestors.

Many people seem to have absolute confidence that we can and do know that the world could not have been created a few thousand years ago by God, but rather has existed for incomprehensibly vast ages; and has ever since the beginning functioned, without exception, according to the laws of physics as we know them, without variation or interruption. Perhaps no arguments or facts I might present could shake such faith, but I hope there are some who will be open-minded enough to concede that we can't be sure about the ancient past. At any rate, I'd like to look at a few beliefs about the past that are especially relevant to the creation-evolution debate. I believe there is a big problem with what most people think we know, and how much we don't know.

Chapter Two, Section Three: Why Divinities instead of Darwinism?

The previous two sections of this chapter are only intended to give some perspective to the historical background of the creationism-evolutionism debate, and to raise questions that are seldom asked: Why didn't this debate happen long ago? And, Why did religion and belief in divine creation arise in the first place, rather than belief in evolution?

I want to emphasize the major change vitally connected to these questions. Currently, there is widespread belief in the "fact" of human prehistory lasting hundreds of thousands of years (or millions, starting with the first chipped-stone tools). This is a fairly recent development. Only in the last 200 years have many people thought that humanity or the current world began more than a few thousand years ago. These beliefs are found not only in Middle-Eastern and European traditions, but also Chinese and Western hemisphere cultures. As we've just seen, these accounts are closely matched by the known facts of the beginning of written history, the oldest written stories and extensive accounting, and the sudden and relatively recent appearance of civilizations all over the world.

If the point has been lost in all the words and large numbers, perhaps a more graphical view would help. Rather than go over all that supposed prehistoric time again, I'll leave it to you for an exercise: Take a small symbol, object, or measured line segment to represent the 5,000 years or so we know from the records of human witnesses, and make a line of them to represent the 2 million or so years that our genus is believed to have been in existence, or the roughly 300,000 years that our species is supposed to have been around. Ask yourself if so much has been produced, built up and sometimes left behind in just the last bit, why did so little happen for all that time before?

Early evolutionists thought that Neanderthals and other human fossil species they discovered were probably much closer to apes than the evidence we now have indicates. We now know we can't chalk up the lack of earlier advancement to ancient humans being so unintelligent. If it seems a stretch to believe that humans lived for such a long time with such slow advancement technologically and culturally (and it should), there is less reason to believe stories about the development of religious belief. Many people seem to regard it as fact (and it has been presented in textbooks and educational TV shows) that primitive human ancestors made up stories about ghosts, and powerful beings controlling the forces of nature. These beliefs about a prehistoric time are based on little more than atheistic speculations, expectations, and faith. Theories about the evolution of religious belief therefore indicate a psychological need to explain why, in an absolutely natural universe, religion exists at all. In other words, the very existence and global occurrence of religious belief is an inconvenient truth for those who wish to believe that the natural world is the sum total of reality.

It is outside the scope of this book to defend the specifics of Genesis and the rest of the Bible, but in general, one might ask why neither they nor any other ancient writings seem to support the generally-accepted prehistory of man. Some ancient writings refer to legendary dynasties going back longer than 6,000 years (Plato set the myth of Atlantis about 12,000 years ago), but there's no trace of memory or tradition pointing to tens of thousands of years. There are many stories of a great Flood, but no echoes of surviving the harsh conditions of the Ice Age for untold years.

Look at it this way: it easily could have happened that the Scriptures included statements that not even Fundamentalists could support or explain as figures of speech, etc. For example, other ancient writings describe the Sun going through a tunnel in the Earth at night. Scoffers make much of the Biblical reference to the "firmament" of the heavens, but the Bible doesn't describe the stars as lights shining through holes in the sky, nor angels standing on the top side. If evolutionary history were true, vast and advanced civilizations might have developed in Africa, Europe, or China vast ages ago, unknown to Moses or later scribes. Even more dramatically, a separate genus of intelligent mammals might have evolved from New World monkeys and created a civilization in the Americas! But nothing like that has turned up. Instead, there is much evidence that modern agriculture and civilization began in the Middle East, consistent with the Biblical account of the Ark landing in Turkey and people spreading from there.

I would also like to suggest that there are unique aspects of the Bible which should be given more consideration than its detractors are willing to do. The scoffers mock such things as divine creation in six days and "the talking snake" in Genesis, but the human tendency seems to be toward far more elaborate and bizarre accounts. Other ancient religions include a panoply of fantastic gods, giants and monsters, from whose bodies and/or fluids the Earth is made. Many ancient mythologies include whole tribes of sky-scraping giants or bizarre monsters encountered by heroes in the wild. Genesis contains none of these popular stories.

Then, too, there is the question of the monotheistic nature of Genesis. The modern explanation is that this is a cultural advancement, a stage from polytheism to deism or atheism. There doesn't seem to be much more to that claim, however, than wishful thinking on the part of deists and atheists. The Jews, and later on the Christians, arguably took their belief in God more seriously than did the believers in all the many religions which have vanished long ago. If monotheism had been invented because it was a natural step toward atheism, why did other religions continue to have pantheons of many gods?

For every speculative story about the way religion started and evolved, there are many questions that cast serious doubt on it. It all seems beyond testing or discovery, but the Biblical account is still believed by many people, including a number with perfectly good degrees and careers in science, medicine and technology. So let's think about it from the other side: Why didn't humans come up with naturalistic evolution to start with?

Modern people take pride in believing that evolution is a highly advanced, scientific theory that primitive people could never have imagined. Again, this seems to have little, if any, basis. Think about what evolution says about the world, and then look at examples of "primitive" imagination at work in mythologies. Living things changing over generations wouldn't be hard to imagine for people who told many imaginative stories, and created images of people with animal heads. The idea of Darwinian evolution requires only two simple observations and one leap of imagination. The first observation is that in each generation, the offspring have differences from their parents and each other. The second observation is that these differences affect their odds of surviving and having successful offspring of their own. The leap of imagination is that, given enough time and generations, this process of variation and weeding out of differences produced all of the differences in all living things.

Quite a leap, but is there any reason that ancient humans could not have thought it up? Surely the people who figured out how to domesticate and breed wild goats and bulls into useful forms would have made the two observations. Variation in domesticated animals through selective breeding was one of Darwin's main illustrations. People who imagined all sorts of monsters and dreamed up tales of strange transformations could surely imagine that animals could change over time into other animals. Even if belief in the supernatural came first, and gods are merely exaggerated versions of ordinary men, why not surmise that the gods had guided the breeding of some plain little animal to produce all the others?

Early humans had to be intelligent to survive without all the technology we have today, and to start inventing it. They also lived closer to nature and had more nature to observe, before civilization spread and many species went extinct. From cave art to the earliest written stories, we see that our ancestors could think very creatively. They had no problem with believing that nonliving objects could come alive, so surely there was nothing stopping them from coming up with the idea of evolution. If evolution of microbes into all forms of life were true and a vital part of nature, the force which formed ape-like creatures into the early humans, living alongside now-extinct primitive "cousins," wouldn't it have naturally been the star of their stories? Wouldn't it have a greater survival value than stories about non-existent beings? Wouldn't it be more comforting than stories about unseen powers magically causing disasters at their slightest whims? And if evolution didn't occur to them, or didn't provide enough survival value to stick, can it be any more relevant now that our society is so removed from purely natural living?

Evolutionary ideas did crop up once or twice in history. However, they never became very popular until after Darwin's formulation. Thousands of years after the rise of civilization and religion, a couple of the Greek philosophers taught that life arose in the water and land animals were descended from amphibians. This shows that the basic idea of evolution – a gradual development of living things from simpler ones – was not beyond early humans. It just doesn't seem to have provided any benefit, or inspired any advances, or anything else to make it catch on. When science (including biology) did begin its great advance, it did so without any evolutionary theory, and the introduction of evolutionary thought arguably led to several harmful ideas. Even the benefits attributed to the theory of evolution could have happened without believing that evolution produced all the diverse forms of life.

There is one more stone to lay in the foundation of this systematic look at the arguments. We've considered the basics of thinking and "knowing" things, glanced at the messy playing field where people are arguing for and against all sorts of views, and finally looked at the utterly foggy stories about prehistory that cloud the debate. The next chapter will close Part One with some thoughts about the practical limits of knowledge and observation when applied to questions about unobserved origins, explain further why this book focuses on the two extreme views, and look at several things these extremes have in common. This foundation may also serve as the beginning of a new way to look at the subject.

Chapter 3: Making the First Move
(Or: Choose Your Rut Carefully)

Section 1: Square One Is Before and Beyond Science
(A non-scientific explanation of existence is unavoidable)

Each third chapter of this book will look at the application of the relevant philosophical views and scientific studies for the historical period examined previously in that section. In this section, as we've been considering the philosophy and history that preceded modern science, we will look at issues that must be considered before the application of scientific research. While science was designed to avoid philosophical issues as much as possible, science can't avoid having a philosophical framework of its own. In fact, the current philosophical framework is at the heart of the debate.

There's an old comedy routine in which one person tells another that he doesn't know where to start in describing a situation. The other person says "Start at the beginning." The first person agrees and then starts describing when and where he was born, or even how the Earth began. Going so far back is funny for typical situations, but in the big picture in this case, going all the way back is inevitable.

Modern science was named and became organized long after it began to be practiced (as natural philosophy) within the framework of a Christian, creation-based world-view. Significantly, scientific research itself confirms that the question of where the universe came from is beyond science.

The way in which you have arrived at your personal belief about how the universe as a whole came into existence strongly affects the way you will decide everything else in the debate. You may be locked by your personal philosophy into accepting one explanation or another, regardless of known facts and logical arguments, or perhaps you will be left straining to maintain some compromise. You may not be willing to acknowledge there is a valid debate. Still, it is always possible to re-assess core beliefs and change important opinions.

"How does anything exist at all?" is the first and biggest question we must consider. This is the question that all searches for answers to the question, "How did we get here?" ultimately come to. Rather than being forced to consider it eventually, or to try to deny its relevance (as some do), let's begin at the beginning, with square one.

There are two logical possibilities regarding the existence of anything: it always existed, or it is the product of something else. In all of our observations, nothing lasts forever. Considering all the things that have been observed to be produced by something else, we can see they are always made of parts that existed before. So the next step of logic is that either the source of everything we see is beyond our limits of observation and has always existed, or that the tiniest bits that make up everything have always existed. Either way, at the end of our search for the very beginning, we are left with only one possibility, that something has been around forever. It might help to think of it from the negative side: If there ever was a time when nothing at all existed, then nothing would ever exist.

One might argue that it is also possible that everything, or at least the first something, popped into existence from nothing (and for no reason, with nothing to cause it). However, if it were possible for that to be the origin of the universe (or of something that could produce the universe), then it would not only be beyond science, but also logic and reason. Even worse, we would have no reason not to expect similar, or at least smaller departures from natural law happening randomly, perhaps rather frequently. An entire universe that began for no reason should not be expected to go on to operate in a consistent, reasonable way throughout its time and space.

It seems a quick preview of the historical sections is called for: For most of history, there was little that could be done to study this question. In some myths it appears that matter always existed along with the gods, if not before them. Some of the Greek philosophers, about midway through recorded history, developed theories that one "element" (or a small set of them, such as fire, air, earth, and water) was the source of all other things. A couple of the philosophers put forward the first version of the atomic theory – the idea that everything is made of many different tiny particles that can't be broken down.

One of the most famous and respected Greek philosophers, Aristotle, thought about how every moving thing we see is moved by something else that was moving. Seeing how illogical was the notion that there was an endless series of moving things, he concluded that something had to start the ball rolling, so to speak. In other words, if there never was something that had the power to start motion without having been moved by something else, then nothing would ever have moved. This "prime mover" or "unmoved mover" was a purely philosophical concept, not the god of any religion.

Then the Grecian empire fell apart, the Romans took over, and philosophy became largely a matter of choosing one of the previous schools of thought and arguing with everyone who chose something else. Also in this time, however, Christianity had begun. It spread through the Roman empire, taking with it faith in One God, the YHWH of the Hebrews. YHWH (later expanded and Latinized to Jehovah), signifies "I AM THAT I AM," in other words, the One Who exists. Christians came to see this divine revelation as including the limited truth that Aristotle had derived later by logic. This belief provided a world-view with a non-created Creator Who established natural law and did not whimsically or randomly interrupt its normal order at the whim of base, petty desires – as the lustful, squabbling, pagan gods did in many tales.

Things began to get especially interesting after Christianity had spread throughout Europe and then re-discovered more works of the Greek philosophers. Arguably the joining of the two different cultures was the major factor in igniting the cultural revolution brought about by scientific studies.

Section 2: Objections to Considering Divine Creation

We'll look at details of how the supernatural became something to ignore completely in a later chapter, but I want to emphasize now that while similar ideas had been around since before the time of Christ, they never became widely popular until centuries after the beginning of the scientific revolution. Science began with the understanding that the natural world it studied was surrounded in time, space, and an ethereal plane of existence, by a greater, supernatural realm which upheld and could interact with the natural. Eventually, the successes of explaining things currently observed in nature in terms of natural forces led some people to explain the unobserved origins of things under the assumption that natural forces are all that are needed to explain everything. At first this was done in the belief that God would never alter or violate a law of nature after creating it. In the end, this line of thinking injected into science (at least in some areas) what amounts to applied atheism: the assumption that nothing beyond Nature ever significantly affected anything in the universe.

It may be comforting to think that we don't have to be concerned about any God, gods, demons or anything else that might surprise us. With the rise of the deistic and naturalistic views, some people began to believe that matter and energy had always existed. They saw no need for an active Creator God to create matter or start it moving. They asked those who believed in the Creator, "Who created God?" By the 1960s, a number of people boldly proclaimed, "God is dead."

However, what I think of as the "neat and tidy" or "everything is simple and ordinary" view of the universe had begun to unravel early in the 1900s, if not earlier. We began to discover that nature itself is stranger than we thought possible, and has not always existed as we know it. On the smallest scale, matter is now known to be much more complex and mysterious than a collection of solid atoms. Likewise, energy is no longer considered to be simply matter in motion, or pure force. Matter/energy is no longer believed to have existed eternally; the entire universe is believed to have started in the Big Bang, less than 14 billion years ago. This is an incomprehensible stretch of time, but it is not infinite. One can ask "What came before the Big Bang?" and arguably there can be no scientific answer. The question itself is unscientific, yet the modern practice of science has led us to it. It is illogical to proceed with scientific endeavors without acknowledging that science is not self-sufficient and all-encompassing.

So philosophical questions have acquired new life in new forms: What started the universe? Another universe? An infinite series of universes? What caused the Big Bang? Is there a reason the universe (especially Earth) is "just right" for living things to exist? At some point, naturalistic answers to such questions must be reduced to nothing more than atheistic equivalents of "God did it." At any rate, it should be clear that if science is going to dive into questions of ultimate origins, it is arbitrary and unreasonable to exclude consideration of supernatural creation.

Scientific investigations began in earnest at the same time as the Reformation, and continued within a framework that included divine creation for hundreds of years. To the scientists in the late 1800s who believed science had built its own new framework of simple natural laws alone, the theories of modern physics and cosmology would be shocking and bizarre. A couple of the major components of Darwin's theory were discarded almost immediately, and the sufficiency of mutations and natural selection is now doubted by some evolutionists. Furthermore, the philosophy of science which accommodates divine creation has been regaining strength since the 1960's. The scientific effort to understand our world would be strengthened by giving this approach free rein to open up new lines of investigation.

So, there is no basis to the common objection that allowing consideration of anything supernatural in scientific investigations would lead to every mystery being treated as an inexplicable act of God. Science began with belief in God's supernatural creation and sustaining of the universe, a belief which remains a viable and logical position. Likewise, there is no foundation for the argument that we couldn't rely on our scientific knowledge if we allowed for divine creation, as God might change the laws of physics on a whim. We might argue, if the universe began because of some unknown and possibly random natural process not subject to current natural laws, our assumption that natural laws will remain constant would be even less logical.

A common objection to considering divine creation is that there are too many possibilities. It's said that if we "teach Genesis in science classes," we'd have to do the same with all religious creation accounts, which supposedly is undesirable or impractical.

This is only a scarecrow argument; there is no real danger. Historically, only biblical creationism was involved in the origin of modern science, and no other religious beliefs have played a large role since then. Some religions seem to have never claimed that the supernatural realm ever affected the natural world in a way relevant to scientific studies. Besides, if some of them do make such claims, then why not let them present their views, too? However, since all positions which include supernatural origins merely suggest possible limits to scientific investigation, the major points are likely to be the same for all forms of creation-based world-views.

Section 3: Why Focus on the Ends of the Spectrum?

Subsection A: The playing field is sloped towards the ends

Some people try to avoid any conflict by supporting one of the intermediate or combination positions, such as theistic evolutionism: "God created using evolution." They may object to my presenting issues as if the only possibilities are atheistic evolutionism or Young-Earth Creationism. However, I believe I have good reason to do so, beyond the fact that I am most familiar with my own YEC position, and most concerned about the atheistic evolutionism at the other end. It's certainly not the purpose of this book to bring all creationists to the YEC view. When I do deal with the age question, it is mainly as it relates to the assumptions leading to belief in evolutionism.

Emotionally and socially, a compromise or intermediate position can be very comfortable, but only as a way to avoid the unpleasantness of dealing with contradictory data and debate. During active debate, the playing field seems to have a peak in the center with slippery slopes towards both sides. Theistic evolutionists happily slide over and show solidarity with atheists against the creationists. It seems quite a few others in the middle tend to slip towards that end of the field eventually. It's not easy to take a position that's attacked by the majority of scientific authorities.

If it can be shown that naturalistic explanations are not unquestionable facts, as they are often presented to the general public, then anyone open to belief in God should be willing to move toward the young-Earth creationist position, if not all the way. The flip-side of this is that, if the naturalistic explanations do provide the true, entire history of the universe, then there would be little reason to give much credence to the Bible at all, since in that case it begins with (at best) a strangely garbled, metaphorical account of the beginning, and key parts of the rest which are based on or refer to these parts are weakened. It could be argued that other religions, and belief in God in general, would be similarly reduced to mere psychological aids of doubtful use, like believing in Santa Claus.

Similar to the attempt to avoid the issue by compromise is the position that "evolution is a fact, but the specific mechanisms are debatable." However, there is not much room for variation within naturalistic evolution, given the constraints of physics and biology. If evolution of all life from a simple form were possible, it would have to be very slow, gradual, and largely dependent on chance. None of the mechanisms change the underlying framework of natural variations over time. People using this argument almost never allow for an actively theistic form of evolution. The only reason for appealing to other mechanisms is to cling to faith in the power of evolution despite the scientific shortcomings of the currently popular form of the story. Most of the criticisms now aimed at evolutionism would still apply to any reasonable variation.

Subsection B: The extremes may be too far apart for compromise

Those who still feel that some middle ground is the best should not let that keep them from carefully considering the two extreme positions. Although the human mind is capable of compartmentalization or gymnastics that allow for two such incompatible opposites to coexist, you should at least be fully aware of the incompatibilities.

The official statements of the major scientific organizations do not allow for any supernatural influence – not ongoing guidance, not even pre-planned setting of the starting conditions – but only unguided natural forces without purpose or plan. While only a few scientists actively campaign against religion, the consensus message is that scientists should completely forget about God when they are at work. The scientists who still hold the original creation-friendly view of science, however, not only believe that God had something to do with how the universe was set up, but that He actively created all the major forms of life (and later caused a massive resurfacing of the Earth, during the Flood). The established scientific position makes any middle position pointless – any compromise is still barred from the inner sanctum of the scientific establishment. From the Young-Earth Creation end of the field, every step of compromise calls into question the significance of many important parts of the Bible, which makes a truly slippery slope for those who believe in it.

While the debate is often presented as Religion versus Science, the extreme positions are incompatible on the basis of philosophy of science as well as religious issues. The traditional creationist view does not attack science in general, but it takes seriously the generally-acknowledged truth that there are limits to what can be known scientifically. The naturalistic philosophy view means that science alone can tell us pretty much all that can be known about ourselves and our universe, not just at present but its entire past and (in a general way) future. For those who fully believe this, science answers all the big questions about our physical existence, while religion and philosophy are little more than mental games irrational people play to comfort themselves.

So compromise positions generally boil down to the illogical belief that while science tells us everything about the real, physical universe, somehow there are still important things we could never learn scientifically. They let science eat the whole cake, but want to believe there is something more left to keep. Atheists tend to ignore such positions while attacking YEC, because the halfway views tend toward leaving God's existence on the level of "Yes, Virginia, there is a Santa Claus."

Subsection C: The extremes have common points, and the same data

Looking at the range of positions from yet another angle, comparing the ends in contrast to the middle positions, we can find surprising and helpful similarities. It seems that a number of the middle positions hold to some form of relativism or artificial division of reality. For those positions, there's hardly any point in debating or studying the debate – there's no such thing as absolute truth, contradictory things can both be true in their own ways, and therefore there are no irreconcilable differences. In contrast, proponents of the extreme positions are nearly unanimous in the belief that truth is not relative, and that incompatible beliefs about the world can't both (or all) be true. Both purist positions hold that empirical studies of the world can reveal definite and important truths, even though the data gathered can't provide absolute proof regarding the ultimate questions.

In a way, the view that the debate is one of religion opposing science is a middle position. The creationists say there is no battle between religion and science because what can be fully demonstrated scientifically agrees with the Bible, and the story of evolution depends on a false philosophy. Strict evolutionists argue that there can be no debate, because any view that allows for supernatural intervention in the natural world has nothing to do with science. The middle positions accept both the scientific validity of the evolutionary worldview and the relevance of Scripture, and so they put these incompatible views in the same arena. The purists at both ends should (ideally) have a calm discussion of the philosophical nature of science, and perhaps what, if any, practical value or harm is produced by the kind of science that makes claims about past events before records of observations. In practice, everybody jumps into that arena in the middle, and then there's a battle royal that gets bogged down in messy details and ugly rhetoric.

Subsection D: Both are dogmatic, but conversion is possible

The two extremes are also alike in that they are both dogmatic. Usually dogmatism is attributed to the creationist position alone. Evolutionists almost always strongly deny that there is any dogma involved in modern science, including their belief in evolutionism. They claim that there is no naturalistic philosophy upholding current science. According to them, there is only the method of science, which is the study of the natural world. However, a few evolutionists over the years have admitted that in some areas, science is restricted by the dogma of naturalistic philosophy. Others have stated dogmatically that science must be that way.

Don't forget, creationists point out that whenever something can be repeatedly observed or demonstrated to be perfectly natural, it is consistent with an original divine creation, and there is nothing to debate in all those areas of science. In areas where observation and experimentation are impossible, creationists openly allow their religious dogma to hold sway, while evolutionists insist there must always be purely natural explanations.

Allow me to clarify that, for both sides, the dogmatism doesn't mean that there is no room for variations. The evolutionists debate among themselves about various mechanisms and modes of inheritance and natural selection, the degree of regularity in evolutionary processes, and whether or not evolution inevitably produces certain results. In the YEC camp there are different explanations for why we see light from stars that are farther than 6,000 light years, how much variation is possible within kinds of living things, and many variations of Flood geology. Neither extreme is quite as monolithic as they are sometimes portrayed.

Furthermore, despite the inherent dogmatism, they both have had members jump ship, and sometimes go all the way over to the other side. The large numbers of students who go to college as creationists and come out as evolutionists – and atheists – is a well-known problem for churches. I know of a couple of cases of active creationists who became evolutionists because of difficulties with scientific data. I have also collected the stories of dozens of former evolutionists who took a hard look at what they thought they knew, and decided that divine creation best explains how all the major forms of life came to be. In some cases on both sides it was something quite trivial that started the change – for examples: fossilized traces of burrows on one side, a tiny shell passing through several layers on the other. See the books *Persuaded by the Evidence* and *Transformed by the Evidence* edited by Dr. Jerry Bergman and Doug Sharp.

I must emphasize that while the debate is all about the differences of opinion and belief, there are vast areas of science where there is no disagreement, not even between the extremes. Most scientific research isn't involved in the debate at all, and some that has been brought into it can be used by either side. This common ground includes all of the science that has produced all of the practical benefits of technology and medicine. In fact, much of evolutionary science is accepted by YECs as perfectly compatible with their view, except in name. No specific point that is actively disputed would have any effect one way or the other on the rest of science, let alone technology or medicine. Statements to the contrary are merely more rhetorical scarecrows. It is entirely through the powerful religious, philosophical, and social implications of claims about the prehistoric past that the debate has stirred up so much fuss and bother.

Intermission

This concludes our overview of things which logically and chronologically provide a sound foundation for studying the relevant issues. In the first chapter, we considered our limitations in learning about the world around us, which includes limitations in deciding things through debate. We've seen that it's complicated in this case by confusing variations from one position to the next, especially in the definitions of terms. Then there's the problem of the blurred lines between the different teams, in a sort of game where the rules are made up on the fly. In the second chapter, we looked at how the origins of belief in creation, and of human history in general, are lost in the mists of time. In the third chapter we considered that any set of explanations (religious or not) that tries to explain everything will have to begin with something unobservable and beyond anything in our normal, natural world. Finally, we reviewed some reasons to focus on two very different sets of explanations that nonetheless do have some common ground.

As this whole book is intended to be an introduction to the debate for those who haven't already gotten deeply into it, but would like to, the next two parts are also heavy on background material and history. Remember, many people might argue with much of this first step. Many people still believe that the Creation-evolution debate is merely one part of a larger battle between religion and science. As we look at the roots and development of science, I will show basic concepts and historical facts which have been largely downplayed, overlooked, or forgotten, and others that have wrongly been dragged into the debate, and I will argue that the debate is due mostly to distortions of the original scientific precepts.

Part II
A New Game Is Invented: Science

Chapter 4: The Basic Idea: Building Knowledge with Bricks, Not Straw

Step 1: First, Limit and Set Apart Natural Philosophy

Everybody knows that belief in divine creation was around long before Darwin set evolutionism on the road to becoming the ruling dogma of biology. What's often ignored, however, is how long religion and science (as natural philosophy) operated together with hardly any conflict, and many mutually beneficial interactions. And that they still don't really clash. It was after science was shifted to its current philosophical basis that some people promoted the notion that religion was its natural enemy, and belief in active divine creation was portrayed as especially contrary to scientific thinking. The early partnership of science and religion, and the shift in philosophy of science later, are matters of history that have been obscured and distorted by fiction and outright propaganda. Still, there are some facts and concepts that are clear from historical documents.

As we shall see in the next chapter, scientific research started as an alternative to rationalism and authoritarianism, but not as an alternative to religious beliefs. Natural philosophy moved away from philosophical assumptions and rationalizing, based on the authority of earlier philosophers. Instead, it sought to keep only statements that could be confirmed by observations and demonstrations. It also rejected simple inductive empiricism – recording observations of nature without further thought, or assuming that no exceptions would occur if none turned up after many observations. It sought to study aspects of nature that could be demonstrated to occur regularly, making the philosophical assumption that they were natural unnecessary. This traded off the broader scope of philosophical thinking for the more certain progress of the limited study of things that can be repeatedly observed.

Step 2: Science = Observing, Recording, Analyzing, Predicting, Testing, and Repeating

So, what is science? Some may say it's whatever scientists do, but that's a circular definition. If you look in various books or on the web for current definitions and descriptions, you will probably find some similarly vague, confusing examples calculated to include as many areas of research as possible. There's plenty of room for debate on this definition alone, since science took shape before there was a formal organization.

However, there are historical documents from the time that natural philosophy (as it was then called) was taking the form that produced the marvels of modern science. The method they describe was taught as "the scientific method" when I was in school, and may still be presented that way. The extent to which the method is now being downplayed or rejected as the heart of science shows the strength of bias and influence of those who want to support areas of science that have gotten away from it. Those who approve of the shift would say that science has grown beyond its original restrictive limits. At any rate, it is only in these added areas that there is any debate, and the debated things have produced no practical benefits.

Science is a system for learning about the natural world around us in a way that balances and tests what we think we know from all three sources – the empirical information we receive through our senses, the reports of others, and the rational deductions we make. The idea of science is that nothing should be included until it has passed tests in all three areas of learning: repeated observations, written reports from different sources, and logical explanations for the observations, confirmed by further observations and testing. It's been said that everything in science is still open to review, revision or even rejection.

For most of the history of science, for most of science as it is practiced today, and for all of the science that has practical benefits, the system works like this: First you get an idea about something in the world around us. You may have heard it from someone else (a report) or come up with it on your own (using your reason) after seeing or otherwise experiencing something (receiving it directly with your senses). (I'll use "observed," "seen," etc., as shorthand for any information gathered with any of the senses, or detected with the aid of instruments and experiments.)

For a lot of things, science depends on making sure of what you think you saw. One way is to observe it some more to see if your idea was right. Maybe you had an idea of how natural events will cause an event in the future, and so you might just keep watching and see what does happen. Maybe you caught a glimpse of an animal you think is a new kind, or one considered extinct or legendary, and so you need to get a better look, take a picture, get a casting of a footprint, etc. Maybe you thought that a tiny speck jiggling around in water was being knocked around by particles too tiny to be seen, and so you need to look at it with a powerful microscope. One thing is for sure: if you're working scientifically, you don't expect anyone to take your word alone for anything. You publish your ideas and observations, and hope that other people investigate and confirm or disprove your conclusions.

Usually quite a bit more is required than having someone else, or even a lot of other people, agree that they've observed the same thing. As a simple example, if you see something that looks like a twig but you look again and think it is an caterpillar, having other people look and agree may only confirm there is something that could be a twig or an insect. Judging if it is a twig that looks like an insect, or an insect that looks like a twig, may take some more observation and a bit of reasoning. If it starts walking, you can be sure it is not a twig. If someone grabs it, finds it solidly attached to the tree, cuts it open to see the sap and layers of wood, then it's a safe bet it's a twig.

Lots of people have seen and photographed or filmed UFOs, the Loch Ness monster, Bigfoot and similar phenomena, but exactly what they've seen or captured on film is arguable, and often has been shown to be something ordinary or an outright hoax. On the other hand, giant squid were once considered mere legends, but more recently have been studied so thoroughly (dissection of dead bodies, capture of live young, clear pictures and video of live adults) that they are now an accepted part of science. Unfortunately, scientific men don't always accept explanations for observations with open minds. For example, when J. Harlan Bretz proposed that the channeled scablands of Washington state were formed by tremendous flooding, the hypothesis was strongly opposed by the geological establishment for many years.

So, for the most part, science is more than merely observing things and reporting on them. It is more than giving a lot of reasons for an idea, or pointing to a lot of evidence that is consistent with the idea. It is more than gaining the approval of a lot of experts. All of those things are part of science in practice, but the heart of science has always been the requirement for one step above all that. This final step is to show, in a repeatable way, that something is true beyond a shadow of a doubt. It's not a matter of which authorities or how many people believe something, but demonstrating that something is the truth no matter who believes it or doubts it. It is not enough to have sound and consistent reasons, especially if there are things about it that still don't fit or still haven't been observed, or if other reasons that could produce different results might be the true explanation. It's not how often something has been observed, but how well it has been observed, too. Eyewitnesses can be mistaken, pictures and other indirect evidences can be unclear or faked, but (for example) a living specimen in captivity can be seen by all and studied to anyone's satisfaction. If creatures like Bigfoot or lake monsters could be observed regularly by anybody, by going to certain places at certain times, setting out the right bait, etc, they wouldn't be regarded as legends.

Most people think of experiments when they think of science. Not all science is experimental, but every area of science that has produced practical results has also involved experiments or demonstrations. Experiments are designed sets of controlled observations. The aim of an experiment is to create special conditions that limit the possible explanations and produce clear results that show if the idea being tested is true or not. Poorly-designed experiments can produce misleading results, and scientists can draw wrong conclusions from good experiments, but different scientists running a variety of experiments have been very successful in discovering useful facts about the natural world.

So a scientific statement has to be about something that can at the very least be observed by other people in a reliable way (repeatability), and better yet can be tested with experiments (testability) and possibly proven wrong (falsifiability). At least, that's what everyone used to agree was the way science works.

Step 3: NOMA – The true and logical division

For the first hundred years or more that science was developing, there was little, if any, question about science conflicting with religion. Science, as natural philosophy, concentrated on the things in the world around us which can be demonstrated or observed repeatedly and reliably. It didn't concern itself with how everything came to be in the first place. It didn't deny underlying supernatural causes. The traditional Biblical view provided motivation and a solid foundation for this scientific research. Unlike other religions, it has no pantheon of gods or spirits with conflicting whims representing different parts of nature. The Bible speaks with approval of studying and learning about the natural world. It portrays nature as giving information about its Creator. Genesis chapters one and two indicate that animals and plants can be usefully classified into a limited number of distinct groups. The account of the Flood provides a helpful starting point for explaining many geological features, especially the large numbers of fossils found in sedimentary deposits all over the world. Beyond these starting points, however, the details and everyday workings of nature were left to humans to discover.

This arrangement provided religion and science with what is known as NOMA: Non-Overlapping MAgisteria, or separate areas of teaching. This concept was promoted by the late Dr. Stephen J. Gould (in his essay, "Non-overlapping Magisteria," *Natural History* 106, March 1997, pp. 16-22). However, Gould's description denied (or at best ignored) the historical, Biblical position that the Earth was created in 6 days about 6,000 years ago. Worse yet, Gould handed over to science *all* aspects of the natural world, including its past. To be fair, Gould's position had already been adopted by major denominations and religious leaders, including the Pope. All of that came later. For now, we'll look at how most scientific research sticks to the original plan.

For one example of the countless descriptions of the scientific method, I'll point to pages 6 and 7 of *Elements of Biology: A Brief Course for College Students* by Perry D. Srausbaugh, Ph. D., and Bernal R. Weimer, Ph. D., (copyright 1944 by the authors, published by John Wiley and Sons, Inc., sixth printing October 1946). There, the scientific method is also called "the technique of straight thinking." This textbook example starts with observation:

"The procedure consists essentially of the following steps: observation, analysis, and conclusion. The scientist also very often makes use of experimentation. In the early beginnings of the application of the scientific method it usually stopped short of experimentation, which, of course, also involves observation."

Once an idea becomes part of scientific investigation, it has a special name: "This suggestion may then be adopted as a tentative explanation or hypothesis." Then, there's a special name for the records of the observations: "The records of the observations made may be accumulated as notes; they are called data."

Finally, if observations and experiments keep supporting the hypothesis, we come to one of the most controversial special words of all: "If a number of experiments are made the results of which seem to point toward the same conclusion, the hypothesis will seem to be confirmed. This hypothesis is subjected to further experimentation and observation. If these additional experiments and observations continue to square with the hypothesis it may then be designated as a theory."

The controversy is over how well Evolution (as a general idea, evolutionism, as opposed to the precisely defined and scientifically verifiable statement of variability in a population over generations) deserves to be called a theory, with some on the side of Evolution believing it should be considered the sort of theory that is more like a law: "The theory may be subjected to continual examination, and, if substantiated by additional facts — all the facts available — it then becomes a law." The scientific status of the theory of evolution according to these standards depends largely on which definition of evolution one has in mind.

There also seems to be a lot of reluctance on the part of evolutionists to consider that Evolution, like any other idea in science, may still be overturned: "The facts supporting theories or laws are made known to all who care to hear or read. Others may repeat the experiment, and if facts are uncovered which are contrary to the hypothesis or theory it will have to be modified or abandoned. Thus we see that the scientist insists on factual evidence to prove all points." Evolutionists will give enthusiastic lip-service to the idea that, hypothetically, evolutionism could be ruled out by new evidence, but quickly rush to attack, explain away, or make excuses for any missing, inconsistent, or contrary facts.

Creationists point out that there are a number of aspects about Evolution (again, the general concept) which have never been observed (let alone experimentally verified), there are a number of facts which are inconsistent with it, and argue that that two or three major scientific laws strongly suggest that Evolution is impossible.

There is the law of biogenesis – life always comes from life, in every case we've observed. The laws of heredity state that living things inherit pre-existing traits from their parents, and the laws of probability weigh extremely heavily against random changes leading to increasingly complex beneficial new traits even under ideal conditions. While it hasn't been rigidly defined and quantified, scientists have noted the general trend of the universe toward organizational entropy, that is, for things that are complex to fall apart, not become more complex. The evolution of life from microbes (which came from raw chemicals) is supposed to be the single great exception, but how it managed this feat has never been precisely explained and demonstrated to be possible.

Therefore, Evolution does not qualify as a scientific theory. It is a faith that takes the rather trivial technical theory of evolution and leaps between points of data with a filament of imagination to claim far more than can be substantiated. Imagination is important for guiding scientific analysis, but it cannot be substituted in the scientific method in place of observation and demonstration; or clear, precise and thorough proposals of complete cause-and-effect sequences.

In response, evolutionists have emphasized the naturalistic philosophy of the new view of science, and the sheer volume of scientific reports supporting evolution. Evolutionists also use a newer, lower standard for supporting data than is found in other areas of science. Data which don't clearly contradict the theory are played up as being consistent with it, and this accumulation of non-contradictory data is then taken as strong support for the theory. They don't seem to realize that a lot of this data is also consistent with creationist views. Scientists in some fields which produce statements about unobserved past events also downplay the usual requirement of repeatable observation, as well as the gold standard of controlled experimental demonstration. They don't seem to realize that this new view of science undermines the true scientific method. They retain the trappings of science and believe they are practicing it, but they are building a house of straw and then putting shingles six feet out on the fog.

The book I've been quoting, written at a time when evolutionists felt safe in claiming there weren't any scientists who doubted Evolution, still exhibited the need to uphold the belief that Evolution was a perfectly sound scientific theory. Point number two of "Some Misconceptions of the Theory of Evolution" (p. 438), responding to the claim that "Evolution is a religious belief or creed," states: "This is not true. Evolution is a scientific theory based on conclusions drawn from accurate observations and data obtained from experiments. It does not concern itself with the spiritual and metaphysical, nor does it make a choice between religion and science necessary."

Note that it does not say that organisms have been observed to evolve into different kinds of organisms, nor that experiments have demonstrated that it is possible for them to do so. When it comes to Evolution and certain other areas of science, the standards are lower. Furthermore, the proper boundaries of science have been exceeded. The origin of life and its many wonderful forms had always been a matter of religious faith, and no rival account can be completely free of spiritual and metaphysical implications. Darwin and other early evolutionists clearly saw the theological underpinnings and implications of the theory. It should be noted that evolutionists will claim that the origin of life is not part of the theory of evolution, but on the other hand, it has commonly been presented in biology texts, in chapters focusing on evolution, as if it were as factual as anything else.

One thing almost everyone seems to agree on is that a choice between religion and science is not necessary. The disagreement is over the way in which the two can be harmonized.

Missteps: Seeds of Trouble

The philosophy behind the geological and evolutionary alternatives to traditional beliefs about the divine creation of life, the Flood, and the age of the Earth was not developed until most areas of science were well-established. The alternatives didn't pop out of the blue, however. There were several ideas, inherited from the Medieval Scholastics and the ancient Greek philosophers they idolized, that paved the way for naturalistic philosophy to infiltrate science.

One of these ideas was "the Great Chain of Being." This idea appears to be a mix of observation, extrapolation, philosophy, and the religion of Gnosticism. It stated that there was a perfect, complete series of things in existence that had emanated from God down through all forms of life, all the way down to the smallest and simplest form of matter. This is very different from the Biblical description of God purposely creating distinct kinds of plants and animals, but some people adapted this idea to Christian philosophy with the reasoning that the Creator would have created everything possible, without any gaps. There was no room for evolution, since everything existed from the beginning. The general idea also gives a reason to expect at least some intermediate forms which are not parts of evolutionary transitions. On the other hand, the idea that such a complete series existed would have made it easier to imagine that the diversity of life had been produced gradually by such small steps. Perhaps the Great Chain of Being also supported the idea of life evolving from something not alive, by placing some life so far down the ladder as to be little different from non-living things.

The other idea that may have helped people accept evolutionism was this belief in spontaneous generation of life. This is the belief that some living things can be formed directly from non-living things. Belief in spontaneous generation was not entirely unscientific at the time, as it was partly based on observations. For example, maggots seemed to be produced directly from rotting meat (nobody noticed the tiny eggs laid by flies). This idea was soon disproved once the scientific method was applied, but the belief that some very tiny, very simple forms of life arose directly from non-living nutrients remained until after Darwinism had begun to be accepted. While the origin of life has never been a part of the theory of biological evolution, the belief that non-living matter could turn into living things naturally must have made it easier to accept the idea that simple living things could change naturally into more complex ones.

There were two other powerful ideas that helped change science and led to acceptance of evolution. The first of these was that studying nature was like reading a book. At first this was proposed in a way that was entirely compatible with the Christian religion. Arguably, the Biblical idea that studying nature was like hearing a proclamation of the glory of God was a major factor in the beginning of modern science. All of the earliest scientific men believed they were studying the handiwork of God and thus learning about His power and intelligence.

However, the "book of Nature" idea grew out of control and created problems from both religious and scientific perspectives. The problem from the religious side was that elevating the study of nature to the position of a second book of revelation encouraged the view that the opinions of men about nature were equal to the revelations in the Bible. This view became more common as modern religious ideas undermined faith in the Bible, and the pride of some scientists led them to present their opinions as facts.

From the scientific view, the problem was that it downplayed the need for repetition of observations and experiments – when someone reads a book for you, you generally don't feel the need to see for yourself if they read it right. If nature is seen as a sort of book, it's much easier to trust someone who tells you what he thinks it says, and leave it at that. Worse yet, it's easier for scientific experts to make everyone else feel like illiterates who are unable to read the book of nature, and so have no choice but to simply trust what the experts tell them. This downplays the need for independent testing and searching for alternative explanations by other researchers.

The final idea that paved the way for evolutionary views is the fundamental idea behind whatever dispute there is between religion and science. The idea is that Reason is by far the most powerful and reliable means of knowing things. This philosophy, Rationalism, is not so much a specific concept as it is a faith in the power of human reason. There is much support in the Bible for having a great appreciation for reason, and this was reflected in the writings of some prominent Christians even before the scientific revolution. The question is, how far can we trust human reason to interpret the "book" of nature? Rationalism produced the belief that scientific reasoning could reveal the facts about everything, from the indefinite past to the limitless future, with practical difficulties but no theoretical limits. While many religious people would come to hold this view (while believing there is also a supernatural realm), leading atheistic proponents have argued that scientific explanations for everything in nature leave no need and no room for religious beliefs.

Most areas of science still focus on how things work in nature now. Articles in scientific journals regularly call for caution and further observations or experiments before placing confidence in results and conclusions. The twentieth century produced a number of examples in which failure to take such precautions led to serious errors. But the early successes of scientific research encouraged the belief that much more knowledge of nature was possible, even about the unobserved, non-reproducible past. At first, this was done in the belief that the Bible provided a divine revelation of a few major supernatural events that human reason could work around. However, later researchers believed that God worked only through natural means, and reason alone was enough to interpret the entire Book of Nature. Interpreting geologic formations as built up by ordinary natural processes inevitably led to conclusions about the past that were contrary to the Bible. This approach, perhaps largely due to the prestige it gave to its proponents, was seen as modern and progressive.

So, while science began with a requirement for repeated observations and testing, there were cultural and philosophical attitudes that would lead to the inclusion of areas where this wasn't possible. We will focus on the details of philosophical and historical developments that established this newer, expanded conception of science in part three. First, though, let's look at some historical details of the beginning of science, and then consider how the original, focused idea of science continues to be the key to the most reliable and useful results of scientific research.

Chapter 5: The Game Started with One Team

Section 1: Roots

When I was young (and perhaps still for some people), it seemed as if "everyone knew" that the history of science went something like this: The ancient Greeks had invented science and passed it on to the Romans, but the rise of Christianity caused the downfall of the glorious Roman Empire and plunged Europe into the Dark Ages. The Church (which was 100 percent loyal to the Pope until Martin Luther came along) kept everyone in the dark for centuries. The church wanted to keep people ignorant and promoted superstitions, the belief that the Earth was at the center of the universe, and that the Earth was flat like a pancake. When Greek science was re-discovered in the Renaissance, scientific men like Galileo and Columbus rejected religious thinking, so the Church fought against science. Fortunately, the free-thinking, religion-opposing scientific men prevailed. This may have incidentally inspired Martin Luther to start the Reformation, but that was a sideshow on the path to the Age of Reason, also known as the Enlightenment, which finally freed society of religious superstition and allowed Science to fully blossom and bestow its bountiful blessings on everyone.

I remember children's shows and cartoons, as well as history texts in school, reflecting parts of this view. The actual historical facts, however, are quite different.

The Greeks were far from the earliest people to demonstrate advanced use of their intellect. Relatively advanced mathematics, detailed observations of nature, and feats of engineering are known from long before the first great Greek philosophers. Some examples come from the earliest traces of the first known civilizations. Consider the engineering feats of the hanging gardens of Babylon, the pyramids, and the rest of the seven wonders of the ancient world. To some extent, Greek scholarship had faded before the birth of Christ, and the Roman Empire, for all its impressive engineering feats, had never advanced much from where the Greeks left off. It should also be acknowledged that the scientific and technical advancements that were made in Egypt, India, the Middle East, and China from the beginning of history to the beginning of Western science have seldom received their due in Western literature. However, science as we know it didn't develop until after Christianity had been established throughout Europe.

The "Dark Ages" were plagued with plagues and attacks on Europe from Huns and Muslims, but learning continued and some technological advancements and studies of nature continued to be made in Europe here and there, and even increased in some areas. The cathedrals that were built and still stand show excellent engineering. The ideas of a round (spherical) Earth and a Sun-centered system had been known and largely accepted before Galileo and Columbus, who were both devout Christians with no desire to oppose religious beliefs. In fact, there wasn't any opposition to the idea of a round Earth from those who turned down Columbus (they argued correctly that the world was much larger than Columbus estimated, too large for his plan to sail non-stop to China or India). The development of science in general continued for centuries by men who apparently had no more doubts about religion than most other Europeans. Almost all areas of scientific research were well-established before any organized effort by "free-thinkers" to overthrow religion and claim science as their own.

How the truth came to be so badly distorted is part of later history, so I'll discuss it in Part Four. I believe that reviewing some important developments of both science and religion in their historical order may help to give a better picture.

First, let's recall that some of the earliest examples of human civilization, the pyramids, Stonehenge, the ziggurats of Sumer, etc., show signs of remarkable intelligence, architectural and engineering capabilities, and awareness of astronomy. There are indications of relatively advanced concepts of mathematics in ancient Egypt, Sumer and India. Rather than advancing from there, however, it seems that a lot of knowledge was lost over time. Some of it was passed on to, or rediscovered by, the ancient Greeks, but their philosophical preference for pure mathematics and philosophical reasoning caused them to reject irrational numbers and experimentation, both of which are vital to advanced science and practical applications.

The Romans acquired and maintained schools of philosophy and learning pretty much as they found them in Greek culture, and this was later inherited by the Christians and northern European "barbarians" who filled in the vacuum created by the crumbling of the Empire. As the political power shifted, Greek and Latin were studied by fewer and fewer people. Much of the knowledge and philosophy that had been written in those languages was lost to Europeans. But not entirely, and other factors were preparing the way for a new and unprecedented increase in learning and knowledge.

One of the last great "scientific men" of the Roman Empire was Claudius Galen, who studied human anatomy and medical treatments before A.D. 200, while Christianity was still a minor factor in the empire. About 100 years later (A.D. 303-312), the last and greatest Roman attempt to wipe out Christianity began, but it ended suddenly with the conversion of Emperor Constantine, after he claimed to have seen a miraculous apparition. About another 100 years later again, Augustine (354-430) was converted, and demonstrated that his training in classical reasoning and philosophy (especially that of Plato) could be used in a Christian framework.

The Roman Empire, by now officially Christian (arguably a nominal Christianity that had incorporated pagan elements), continued to decline. For another 200 years or so the story is about the northern barbarian tribes taking over the remains of the empire and then fighting among themselves. Finally, however, the ongoing missionary efforts converted the majority to Christianity and gave them some sense of unity. Then for another hundred years or so they were threatened (and further united) by the rise of the Islamic Caliphates and Ottoman Empire, which took over what would later be Spain and made significant advances into Eastern Europe. For about 400 more years, Europe seems to have been frozen in a struggle for mere survival, while in Islamic lands there was prosperity and development (not just preservation) in a number areas of learning, especially mathematics, astronomy, and medicine.

However, even in these "Dark Ages" of Europe there were still centers of learning where some of the knowledge and logic of the Greeks was preserved, largely through the influence of Augustine's writings. Anselm of Canterbury (A.D. 1033-1109) is a notable example of those who continued to study and practice logic and reason. Thus, there was fertile ground ready to receive the seeds of newly acquired ancient writings. This happened after 1100, when combined forces from all over Europe had finally made a major advance into the Islamic territory and gotten as far as Jerusalem. This first Crusade was the only one to have any significant success, but it was enough to change the relationship between Europe and the Islamic countries, open the way for trade between Europe and other lands, and bring into Europe fresh texts and ideas from many lands and times.

Then again, perhaps the First Crusade didn't so much change things as reveal that they had been changing. While the Crusade arguably hadn't done that much damage to the Islamic nations as a whole, and most of the territory gained was eventually lost again, it seems that Christian Europe picked up the torch of philosophy and learning even as the people of Islam were about to turn away from it.

Perhaps the new-found ancient texts and Arabic mathematics (which borrowed from India and other lands) were not so much a torch that set Europe on fire, but sparks that contributed to a new sort of flame. The Islamic centers of learning had not gone on to develop an organized scientific program in spite of having had access to ancient texts for centuries. Furthermore, the Reformation and Renaissance wouldn't be in full bloom for another two centuries. They both had their roots about this time, but apparently it took something more to fan the flames.

While there had always been some people in the church who dissented from various aspects of the pomp and circumstance, fables, superstitions, and pagan traditions that had been incorporated into church practice, one of the first major outbreaks began in 1173 with the Waldensians. Likewise some of the learned men began to reject the practice of accepting everything the ancients wrote, or arguing about which ancient philosopher was correct, and instead tried observing and experiencing things for themselves. Roger Bacon (A.D. 1219-1294) is the most outstanding example of the few men who were already advocating the experimental approach, and he also predicted that human technology would one day produce flying machines and other technical marvels.

However, most of the learned men of the time seem to have been satisfied in their admiration of the ancient Greek philosophers and their attempts to increase knowledge by logical (rational) manipulations of statements and "facts" received from the ancient authorities. The works of Plato became favored, and those of Aristotle were banned, largely due to the influence of Augustine's work of 800 years earlier. However, Thomas Aquinas then repeated Augustine's application of Greek philosophy to Christianity, but based on Aristotle's works. Plato had advocated a Sun-centered system of planets, but Aristotle had put the Earth at the center. Aquinas' impressive system of rationalized theology (the *Summa Theologica*) included this geocentric theory. It became part of the academic system and thus a minor point of church doctrine as well.

However, the idea of a flat Earth never became popular in Christian circles. The idea that it was is a myth that was invented much later. No educated Christian is known to have written in favor of a flat Earth for almost 900 years before this time when the seeds of Reformation and Renaissance were being sown.

About the same time Aquinas was setting up the system that would become the dogma of the universities, other social events were going on that would help to break the grip of such authoritarianism. Increased education was beginning to encourage more people to read the Bible for themselves, and the differences from church doctrines they found led the Inquisition in Toulouse in 1229 to forbid Bible-reading by laymen. Meanwhile, a new social class made of tradesmen, merchants, and businessmen was forming, growing and reaching out to the rest of the world. Along with the merchandise and commerce came new ideas. The most famous example of this is the voyage of Marco Polo to China and back, A.D. 1271-1295. The Chinese, like the Islamic nations, had developed ideas and technologies unheard of in Europe, such as gunpowder, paper, and printing, although they had not developed them to the extent that Europeans would.

Section 2: Renaissance and Reformation

Historical periods don't begin and end with signs to announce them, but the Renaissance has been dated between 1300 and 1500 A.D. While classical Greek (and Roman) studies continued and grew, with more texts being re-discovered, there seems to have been little scientific development other than the practical, technological improvements that had been going on quietly for some time. Most of the flourishing of culture that gave the period its name seems to have been in the arts.

However, there were also major developments in the religious realm. While the Reformation is usually dated to Martin Luther's posting of his 95 theses (a list of things that he believed needed to be reformed) in 1517, the work of reforming the church had begun with men such as John Wycliffe and Jan Hus in the last half of the 1300s and early 1400s. Hus had been burned at the stake as a heretic in 1415, but by 1458, a follower of Hus was King of Bohemia.

So at least part of the preparation for the flowering of exploration, discovery, and scientific research was a fresh spirit of religious zeal. While the Inquisition was formed to suppress what the Church of Rome saw as heresy, on both sides the religious beliefs were made stronger than ever. Meanwhile, as trade and technical developments increased wealth in general and the middle class took shape and grew, opportunities for more people to learn more things also increased. Arabic math was replacing the Greco-Roman system. Gutenberg's system of mechanical printing (1439) played an important role in both the Reformation and the scientific revolution. As early as 1466, Bibles were being printed in the common language of the people around Strasbourg.

We can see how dramatic were the changes that began in the late 1400s by the names of some people born between 1451 and 1484: Christopher Columbus, Leonardo da Vinci, Amerigo Vespucci, Erasmus of Rotterdam, Nicolaus Copernicus, Martin Luther, and Ulrich Zwingli. Between them, these men would begin some of the early modern-style studies of nature, begin the European exploration of the Western Hemisphere, begin the Reformation, and revive the theory that the Earth travels around the Sun along with the other planets.

Apparently, the growing changes in the religious and scientific realms had something in common. Both of them were based on a new attitude: that individuals had the right to doubt authorities and think for themselves. In their religion, people were studying the Bible and discovering that some practices and doctrines in their church were not supported by it. In studying how natural things worked, they studied rival schools of philosophy and observed nature for themselves, discovering that the ancients had differences of opinions (none of which were right in some cases), and that perfectly logical conclusions could be very wrong, even when based on much supportive data.

Leonardo had to see for himself how the human body was put together, how glass lenses affected light, how birds used their wings. Columbus had to see for himself if he could sail across the Atlantic and reach China. He was wrong, but fortunately there was land much closer than China. His successful failure proved the value of trying an experiment in spite of perfectly logical and correct reasons for not bothering.

In a similar spirit, Martin Luther studied the Scriptures for himself, and later pleaded with those who opposed him to simply show where he was wrong according to the Bible and sound reasoning. Calvin wrote of the value of reason when properly applied. The reformers of religion and the explorers of nature both insisted that reason be tested against something solid, and that the pronouncements of human authorities be compared to foundational information open to all to see.

Probably no one factor alone is responsible for the burst of scientific research that began in the 1500s. Previous improvements in technology, increased wealth that was spread to more people, commercial and cultural exchanges with distant lands, the ability to make more copies of books with less effort and expense, new ideas in mathematics, challenges to the monopoly of the Church of Rome – all of these and perhaps others had been developing for some time, and at last came together in full force. But for the purpose of this book, it's vital to note some things that have been definitely established by historical documents:

- There's no indication that a weakening of religious belief was part of the recipe.
- There's no historical documentation of belief in a flat Earth by any church leader of the time
- The heliocentric theory didn't, at first, draw significant resistance (even from the Pope who learned of it)
- Natural philosophy (what we now call science) blossomed for at least a century in an environment where everyone believed that the Earth had been divinely created only a few thousand years ago and had been destroyed (on the surface) by a cataclysmic global Flood. The charge that religion naturally opposes science, or that creationism is a science-destroyer, is obviously false. It was only after another three centuries (early 1800s) that Flood geology was entirely banished from mainstream scientific literature, and decades more (late 1800s) before Evolutionism took over in biology.

In case there's any question about what I mean by scientific studies blossoming at this time, consider this sampling extracted from *The Timetables of History* (The New Third Revised Edition), by Bernard Grun, based upon Werner Stein's *Kulturfahrplan*, A Touchstone Book, Published by Simon & Schuster, New York, 1991.

1524 *Cosmographia*, first textbook on theoretical geography.

1530 First treatise on mineralogy.

1542 Andreas Vesalius publishes first "modern anatomy" book

1544 Georg Agricola initiates the study of physical geology.

1546 Mercator states that the Earth has a magnetic pole.

1550 Siegmund von Herbstein: *De natura fossilium*

1551 Konrad von Gesner: *Historia animalium*, modern zoology.

1555 Pierre Belon: *L'Histoire de la nature des oyseaux* (Literally, "The History of the nature of birds").

1559 Realdo Colombo describes position and posture of human embryo

1560 First scientific society founded at Naples.

1565 Royal College of Physicians, London, empowered to carry out human dissections. Bernardino Telesio (1508-1588): *De rerum natura*, foreshadowing empirical methods of science.

1569 Tycho Brahe begins at Augsburg construction of a 19-foot quadrant and a celestial globe, five feet in diameter.

1576 Clusius publishes his treatise on flowers of Spain and Portugal; beginning of modern botany.

1581 Galileo discovers that a pendulum's swing time stays the same as long as its length stays the same. This will lead to more accurate timekeeping, an important factor in several areas of scientific research.

1583 Joseph Justus Scaliger: *Opus de emendatione temporum* , foundation of modern chronology.

1590 Galileo publishes *De Motu*, "description of experiments on dropping of various bodies."

1599 Ulissi Aldrovandi, Ital. naturalist... publishes his studies in ornithology.

1600 William Gilbert: *De Magnete*, treatise on magnetism and electricity.

1602 Tycho Brahe: *Astronomia Instauratae progymnasmata* gives plans of 777 fixed stars (posth., ed. by Johann Kepler). Galileo investigates laws of gravitation and oscillation (-1604).

1604 Johann Kepler: *Optics*.

1608 Dutch scientist Johann Lippershey invents the telescope. Galileo constructs an astronomical telescope.

1610 Jean Beguin: *Tyrocinium chymicum*, first textbook on chemistry. Galileo observes Jupiter's satellites... Thomas Harriott discovers sunspots. Nicolas Pieresc...discovers Orion nebula.

1611 Maraco de Dominis...publishes scientific explanation of rainbow.

1618 Kepler: *Harmonices mundi*, stating the third law of planetary motion.

1620 Francis Bacon's ... *Novum Organum Scientiarum.*

1627 *New Atlantis* by Francis Bacon, published posthumously. It described a fictional scientific utopia, providing "plans for a national Museum of science and art."

1643 Torricelli...invents the barometer.

But don't forget that religious belief ran strong throughout this period and long afterwards. Both Catholic and Protestant countries produced these great scientists. Most history textbooks put religion and science together only for the Galileo affair, but consider these items from the same period, taken again from Grun's work and also from *The Faces of Origins: A Historical Survey of the Underlying Assumptions from the Early Church to Postmodernism* by David Herbert, M.A., M. Div., Ed. D.; D & I Herbert Publishing, London, Ontario, 2004:

1532 Reformation in France (Grun)

1535 Luther began his *Lectures on Genesis*, treating the creation account as historical, literal, with creation accomplished in "six natural days." (Herbert)

1536 Calvin's *The Institutes of the Christian Religion* (Herbert)

1536 William Tyndale, Eng. reformer, burned at the stake. (Grun)

1541 John Knox leads Calvinist Reformation in Scotland. (Grun)

1543 First Protestants burned at the stake by Spanish Inquisition. (Grun)

1545 Council of Trent meets to discuss Reformation and Counter Reformation (-1564) (Grun)

1560 English Protestants in Geneva produce the Geneva Bible. (Herbert)

1563 Foxe's *Book of Martyrs* (Grun)

1564-1633 – List of commentaries on Genesis shows, "To a man, there was absolutely no doubt that the Bible...was the only trustworthy record of Earth's six-thousand-year history." (Herbert)

1565 Jacobus Anoncio: *Stratagemata Satanae*, advocating religious toleration. (Grun)

1578 Guillaume de Salluste (1544-1590), French Huguenot, publishes *La Semaine* – epic poem on the creation week. Here's a sample: "World not Eternall, nor by Chance compos'd; But of meere Nothing God it Essence gave." It was "possibly the most popular poem in Europe during the sixteenth and seventeenth centuries – was translated into eight different languages." (Herbert) (The Huguenots were a Protestant group)

1611 Authorized version of the Holy Bible-'King James Bible'-published. (Grun)

It wasn't until the end of this period, in the 1600s, that major works began to question or attempt to revise belief in a catastrophic global Flood or the Bible in general. Significant attacks on belief in divine Creation, organized religion, or belief in God came much later.

Herbert notes that the publication in 1614 of *History of the World* by Sir Walter Raleigh showed that: "Ralegh [sic] had full confidence in the Bible (the Geneva version) as a trustworthy and reliable source." He accepted a relatively recent divine creation and the global Flood, but proposed "that the flood had been placid." His reason for this was simply that the precise description of the location of Eden given in the Bible must mean that it had been preserved. This reasoning ignores several statements in the Bible, and foreshadows the inability of later students of geology to properly account for the data within their conception of the biblical framework.

But the first major work departing in a large degree from general Christian doctrine came in 1624: Lord Herbert of Cherbury's *De veritate*, which formed the "foundation of theory of Eng. deism." (Grun) The deists believed in God and creation in a general way, but doubted details of Christianity, such as the divine inspiration of the Bible and the reality of miracles. The idea of rejecting belief in God entirely and exalting human reason to a godlike position wouldn't come to power for another 100 years or more. Its main root has been traced to 1644, with the publication of Descartes' *Principia Philosophicae* and its famous phrase, "*Cogito, ergo sum*" (I think, therefore I exist).

Descartes himself was at least a deist, and included in his work arguments (at least for the sake of appearance) for the existence of God. The strong Rationalist or naturalistic philosophy remained a minority position in science for a long time. Most leading scientists continued to hold mostly traditional religious views until the end of the 1800s. I will look at the effect on science of these new ideas in Part III. For now, I want to focus on three key figures involved in the foundation of modern science: Galileo, Francis Bacon, and Isaac Newton.

Section 3: Three Key Misunderstood Men

1) Galileo Galilei

Galileo has been transformed into an iconic martyr in "the war of religion against science," but this seriously misrepresents the case. As scholarly studies have shown, there are many considerations contrary to this popular view.

There was very little opposition to the heliocentric theory when Copernicus published his work at the end of his life. Grun does note that in 1549, Melanchthon (who sided with Martin Luther) objected "to the theories of Copernicus." However, others have noted that the Pope at the time knew of the work and approved of it, or at least didn't disapprove of it. At any rate, it doesn't seem to have been a major issue or seen as an incompatibility between science and religious doctrine.

You may be surprised to learn that when Galileo first presented his observations and arguments in support of the heliocentric system, there still wasn't any opposition. It appears that the antagonism towards Galileo and his proposals was due first and foremost to personal issues and the power politics in the upper ranks of the universities. Some of the leaders of the Scholastic system regarded certain teachings of the ancient Greek philosophers with nearly the same reverence as they did the Bible. When this "old guard" found it couldn't keep Galileo quiet with the usual techniques, they stirred up religious opposition where hardly anyone had seen any need before. No doctrines were tied to the question of the motion of the planets, there are no verses which clearly state that the Sun goes around the Earth, etc. This is in contrast with other ancient writings, which did make such plain statements. For example, the Epic of Gilgamesh states that the Sun travels through a tunnel in the Earth in order to return to the East.

At any rate, it doesn't make sense to harp on religious rejection of heliocentricity, when Johann Kepler was working on the theory at the same time as Galileo, but was never pressured into denying it. Kepler, working in a Protestant area of Europe and building on the work of Copernicus and Tycho Brahe, worked out the mathematical laws of planetary motion with as much accuracy as possible at the time. Like Galileo, he did not see any incompatibility between his scientific work and his religious beliefs. More significantly perhaps, he did not meet with any serious opposition from religious leaders. The rapid acceptance of his work (in Protestant lands) is an indication that the only incompatibility was with traditions influenced by Ptolemy's theory, not religion or the Bible.

We should also keep in mind that there was no known way to convincingly demonstrate that the Earth goes around the Sun each year and spins around every day. It certainly has always appeared that the Sun travels across the sky. The ancient geocentric system of Ptolemy worked fine for the most part, having been revised and adjusted so that it could be used to predict future astronomical events such as eclipses. If there had been any strong religious or Biblical argument against the heliocentric theory, Christian Europeans might not have considered it the true arrangement until centuries later.

The mathematical description of the heliocentric system devised by Copernicus (possibly with help from Arabic and other texts), supported by Galileo's observations, and developed and improved by Kepler, was simpler and easier to use. Compared to the complicated loops of the old system, like a dance of the ancient gods that the planets were named after, the heliocentric system also seems more like the craftsmanship of a single God who created everything and established their motions with His laws. Kepler felt that his studies were a step in following the thoughts of the Creator, "Thinking God's thoughts after Him."

It is true that the Roman Catholic Church did require Galileo to renounce the theory of heliocentricity, and continued to oppose the theory long afterwards. Galileo was not officially vindicated by a Pope for over three centuries, although opposition to his theory had been dropped much earlier. However, this was an insignificant incident compared to the terrors of the Inquisition, and the wars and persecution back-and-forth between the Roman Catholic church and various Protestant groups and countries. The negative treatment of Galileo and the banning of books promoting the heliocentric theory merely foreshadowed the distancing of most "Scientific men" from established religion.

2) Sir Francis Bacon

Francis Bacon was born just a few years before Galileo, but it wasn't until after Galileo's trial that Bacon published his great works on science. Bacon had been raised in what today would be considered a Fundamentalist family, but after an impressive career in politics he was convicted of bribery and forced into retirement. He doesn't seem to have contributed anything new to science, but his writings about how scientific studies should be carried out helped to standardize and organize the work of "natural philosophy" and shape it into the "science" that existed with little change for a century or two, and is still practiced in the vast majority of science.

In works such as *Novum Organum Scientiarum* (1620) and *New Atlantis* (1627, after his death), Francis Bacon set forth his method of learning things about nature, along the way pointing out the flaws of other ways. He wasn't the first to do so, but for the first time a lot of people were ready to agree. Many had already been practicing the same basic approach. Unfortunately, some people have focused on one part or another, and oversimplified Bacon's plans and arguments, either to support their own pet idea, or for some such ulterior motive to praise or criticize Bacon.

On a technical tack, it would be easy for people promoting either the inductive or the deductive approach to quote or criticize Bacon. The inductive method is to collect a lot of specific examples of something (or notes on observations of it), and draw general conclusions. The classic illustration is about seeing crows. After seeing a lot of crows, you will probably conclude that all crows are black, that there are no white crows. On the other hand, the deductive method takes things that seem to be universally true, or statements that seem logically valid, and draws conclusions about specific cases. The classic illustration is: 1) All men are mortal. 2) Socrates is a man. Conclusion: Socrates is mortal.

Bacon seems to have advocated a combination of the two, but also much more. He stressed the importance of gathering a lot of data first (inductive method), and then using reason (deductive method) to guide further observations (rather than jumping to conclusions), then gathering more data and so on – without any end in sight. Thus, he was not a pure empiricist (going only or predominantly on what can be observed) nor a Rationalist (relying on logic above all else). His system described what men like Galileo and Kepler were doing – exploring nature by logically studying data about it, and grounding their logical conclusions in further observations of nature.

It's important to note that this combination was not merely adding observations on one side to reason on the other (or vice-versa). Bacon made it clear that a vital part of the scientific method was using reason to devise series of experiments: observations of things in controlled situations. The watchwords of Baconian science are: Proceed only with extreme caution. Don't trust your senses any more than you have to. Don't build schemes of logic upon logic. Test, test, test, and test again, as often and in as many ways as you can, to explore or eliminate every possibility you can think of. There was one more benchmark that Bacon insisted scientific studies should (eventually) reach, and although it seems to have been largely forgotten by historians and philosophers of science, it seems to me that Bacon considered it very important. First, however, I feel the need to emphasize another point about Bacon.

Bacon clearly believed in God, found atheism and "free thinking" ridiculous, had a high regard for the Bible, and believed in the divine creation of the heavens and the Earth in six days. Yet here again we find that people with opposite beliefs have found things in his writings to wave as banners for their views.

Unlike some later great thinkers (such as Descartes) who are suspected of having devised arguments for the existence of God merely to avoid being persecuted for the atheistic beliefs their philosophies are said to have supported, Bacon wrote against atheism in clear and strong terms. For him, the evidence in ordinary nature for the existence of a rational Creator was so plain to see that neither miracle nor logical reasoning should be required. Still, he went on to give several strong arguments against atheism. He also disapproved of those people who claimed to be merely skeptical or agnostic, not holding any strong beliefs.

However, he did write strongly against superstition as well, and to many people of "modern" mind-sets all religion is superstition. Furthermore, Bacon wrote against those who built elaborate systems of natural philosophy (science) based on the creation account in Genesis and some other books of the Bible. Anti-creationists take both of these points to place Bacon on their side, especially with the straw-man argument of "the Bible is not a science text."

However, it is clear from yet other passages that Bacon considered science (natural philosophy) to be only a secondary ward against superstition, after the Bible. The study of nature was not for him entirely separate from religion, but was a God-given aid to religion. While the Bible spells out the will of God, the study of nature merely shows the creative power of God. Bacon also believed that the combination of increased exploration, travel and commerce, along with these first steps of scientific research, showed the beginning of the literal fulfillment of a prophecy in the Book of Daniel: "Many shall go to and fro, and knowledge shall be increased." Bacon wrote against undue reverence for the philosophical authorities of antiquity, but never suggested that anything in the Bible might be contradicted. The research society described in his *New Atlantis* was named after Solomon, and was also called "the College of the Six Days Works" after the Biblical account of creation.

On the subject of the origin of the Earth, Bacon made it clear several times that he modeled his approach after the example of God's creation as recorded in Genesis. God created light first, and Bacon proposed to begin with research that was illuminating even if it didn't seem it would have any practical application. God created with a deliberate plan and order, and Bacon took that as a model for his plan to study creation with an ordered system. He predicted many amazing things would eventually be accomplished by science, but he never suggested that science might reveal anything about the origin of the Earth.

Bacon can be cited by advocates of both "pure" and applied science, as he advocates both. It's important to note, however, that while Bacon stressed the importance of starting with observations and experiments that might only provide information with no obvious practical applications, he was arguably still more emphatic that science should be aimed at producing practical benefits. This is his ultimate benchmark. He cited the lack of useful results as a major condemnation of previous academic and philosophical endeavors. He made the eventual fruitfulness of his system the test of its validity. This is consistent with the Biblical principle of "by their fruits ye shall know them" and with God's granting dominion over the Earth to Man in Genesis. I also argue that it is a condemnation of several fields of research claiming to be science, especially evolutionary biology. Several off-shoots of the theory have been harmful, and the few benefits that have been claimed are impractical, arguable, or did not require the theory of evolution. For example, evolutionists claim their theory explains why organisms we wish to destroy developed resistance to pesticides, antibiotics, etc, but the theory did not lead evolutionists to predict the phenomenon and thus prevent it. Furthermore, the variations involved are perfectly compatible with creation-based science.

In his fictional realm of New Atlantis, Bacon described fruits of the Atlantean science program that were clearly intended to serve as predictions of the results real scientific research might one day achieve. It is perhaps significant to note that, like Roger Bacon before him, Francis correctly predicted that human-carrying vehicles would be able to travel under water or through the air. Along with other generally accurate predictions (such as the ability to store and transmit sounds and light), he also made some that now seem amazingly wrong. One was the invention of perpetual motion machines. Another was the ability to generate different forms of living things by properly mixing and heating various soils, gasses, etc. While we now laugh at belief in spontaneous generation, and our textbooks may tell us that Redi put an end to such ideas with his experiments, belief in some form of spontaneous generation (of microbes, if nothing else) continued among professional scientists until the closing years of the 1800s. A form of it continues today, in the belief that somehow the very first forms of life spontaneously evolved from chemicals and energy (abiogenesis). The ghost of Bacon points an accusing finger at this idea, remaining fruitless through all the research from the 1800s to the present – about 200 years of failure. This includes decades of modern experiments, guided by the theory of evolution, attempting to reproduce the origin of life within intelligently designed lab equipment.

3) Newton

Isaac Newton was born much later than these other men, in 1642, the same year that Galileo died. Deism had already been founded and Renee Descartes was about to begin the Rationalist school of thought with the publication of his *Principia Philosophicae* in 1644. By the time Newton published his own masterpiece, *Philosophiae naturalis principia mathematica* (*Mathematical Principles of Natural Philosophy*) in 1687, the so-called Enlightenment or Age of Reason was in full swing. Newton's work, like that of Galileo and Francis Bacon, would be championed by Deists and atheists as supporting their views. This was especially true in France, where Newton's work was translated by the Rationalist Voltaire and one of his lovers. To those who were looking to explain everything apart from God, Newton's laws of physics seemed to offer a complete natural solution.

Newton himself, however, was much more in harmony with the stronger religious beliefs which had been practically everyone's (in Europe). Although his private writings show he had some unorthodox ideas, he showed no doubt about God, His creation of the universe in six days thousands of years ago, and the infallibility of the Bible. The whole period was one of turmoil, change, and mixtures of ideas, but there wasn't even one country governed by men calling themselves Rationalists until the late 1700s. It was in these mid-1600s that Lightfoot and Ussher calculated the Earth's age to be almost 6,000 years. Newton studied and wrote about the Bible extensively. He took the prophecies seriously, and from his study determined when the world (as we know it) would end. He might not have been far off, as we still have a few decades to go.

While it may be argued that Newton didn't see (or refused to acknowledge) the atheistic implications of his own scientific work, such second-guessing cannot serve as a valid argument. Newton, like Galileo, Bacon, and many other founders of science, saw no conflict between scientific research and belief in the Bible, including acceptance of God as the Creator of heaven and Earth and the Bible's account of the global flood. Other ideas were beginning to creep into society, but it would be many years before they had significant impact on science.

Chapter 6: There's Still More Calm than Storm

According to the system of this book, the last chapter of each part is supposed to look at the different interpretations of the scientific data given by creationists and evolutionists. In this case, however, there is no difference about the science involved at the start. For those living at the time of the beginning of science, it would seem strange that some people today see religion as the mortal enemy of science and learning. From the beginning of civilization there was significant knowledge of astronomy, geometry and mathematics, and these were often directly related to religious beliefs and practices. Although a form of evolutionism was proposed in ancient times, it never caught on until well after scientific research began. The virtual marriage of learning and religion continued to be the case wherever such knowledge was preserved or further developed – in Babylon, India, China, etc. As we've just seen, all of the fundamentals of modern science were established by men who embraced the belief that Earth had been divinely created thousands of years ago, and at one time had been covered by a global Flood.

Some might argue that this only shows that it was possible for scientists to hold such religious beliefs in those days before we learned so much more. Yet scientists still can (and some do) hold the same beliefs about the Biblical history of the world. The fact is, the vast majority of scientific research doesn't come close to being involved in the debate, and so is just as compatible with faith in the Bible as it was then.

In geology and related areas, the main area of contention is over the age of the Earth, and in paleontology the hypothetical/imaginary evolutionary lines of universal common descent. Many in the ID movement and various creationist camps accept most, if not all, of those points, too. However, it is these areas that have been plagued with the greatest revisions and the most embarrassing mistakes, such as Piltdown man. The identification and classification of rocks, minerals, fossils, strata, and various formations aren't disputed even by Young-Earth Creationists, and other creationists accept the dates assigned to them. Practical applications, such as the search for fossils and useful deposits of minerals, metals, oil, and natural gas can be done without reference to evolution or dating techniques. All that's needed is recognizing the layers, formations, and types of rock.

Physics and its applications in technology and engineering are almost entirely out of the picture. The main exceptions are the Second Law of Thermodynamics (which creationists argue favors their view), and the use of radioactive decay rates for dating some kinds of rocks (which is used to argue for the Earth being billions of years old).

The predictions of early practitioners of the scientific method concerning machines that can travel underwater, in the air, and to the Moon have been fulfilled. Likewise, the ability to store sound, images, and power, and to transmit them over great distances was foreseen but fulfilled in unforeseen ways. Nobody in those days predicted machines with the capabilities of modern computers. Still, in all these advancements and more, the great progress was made with repeated observations and much experimentation, with all of the major early steps made while creationism was still the only game around. We've learned a lot since Newton's time. Although his formulas are still valid for the common events they were written to describe, we've learned that nature acts differently in the extreme conditions, with the extremely precise measurements covered by Einstein's equations, and that it acts in other odd ways in the subatomic realms described by quantum mechanics. It could be argued that the strangeness of advanced physics indicates we are approaching the boundary between the natural and supernatural.

While evolutionists love to quote the slogan, "Nothing in biology makes sense except in the light of evolution," very many biological studies don't have any references to evolution, and those that do often seem to be merely adding some comments as an afterthought or password. Of the scientific reports specifically about evolution, most of them are about slight changes in populations due to natural selection – the sort of variation that is entirely consistent with a young-Earth creation framework. The vast amount of information that has been gathered about the mind-boggling complexity of living things, even within single cells, favors belief in creation, and certainly was not helped by evolutionism. The field of medicine has suffered several shameful incidents due to evolutionism, but my point here is simply that none of the practical, applicable advances in biology required belief in Evolution rather than Creation.

Likewise, the field of chemistry has come a long way since it first broke away from the secretive mysticism of the alchemists. We understand why different chemicals combine the way they do, and can predict the results of reactions with mathematical precision. No aspect of this progress required belief in evolutionism or billions of years. The one area closely related to the debate, the attempt to reproduce natural conditions and combinations of chemicals that could produce life, has made hardly any progress for over half a century and many different experiments with different approaches. It would be hundreds of years if we counted earlier attempts, made when it was thought that it happened in currently common conditions. There is no disadvantage for a chemist in believing in relatively recent divine Creation, except for the opposition from scientific and academic authoritarians.

If all scientists had kept to the scientific method that has produced so many benefits, there never would have been any evolution/creation debate in science, and all the real, useful advancements in science would have been accomplished just the same. In Part Three, we will look at the new philosophical and speculative method of science that started the trouble.

Part III: Faith in Reason and Nature – The Philosophical Heart of Evolutionism

Chapter 7: Introducing a New Set of Rules.

Section 1: Overview

The previous chapter should have made it clear that science began within a young-Earth creation framework, and that most of science remains compatible with it. Yet today many people regard science and religion as mortal enemies, or at best incompatible. The root of these views goes back long before Darwin, and it began with philosophical and religious men, not scientists. The philosophical invasion of science which turned parts of it against traditional Christianity is tightly bound to its historical roots, from the early 1600s through the late 1700s. The fuzzy lines between the philosophical developments presented in this chapter and the historical developments in science in the next will reflect the infiltration of the naturalistic philosophy into the practice of those areas of science.

The root of the divide between religion and scientific endeavor is pride in human reasoning. The Reformation had begun with reason playing a strong role, and reason also played a major role in the beginning of modern science. However, in the Reformation, reason was always combined with faith in the Bible, and in scientific studies it was combined with observational and experimental verification. The successes of the Reformation and science naturally led to increased confidence in human reason. The increased confidence led to a dependence on Reason to provide "knowledge" which went far beyond actual observations.

It's not hard to understand the attraction of this approach. It comes quite naturally, as everything we experience is filtered or analyzed to some extent by our reason. The kind (or degree) of reasoning that humans do is one of the things which distinguish us from ("other") animals. In the world of pure reason, everything can be neat and tidy, consistent and comprehensive. Or so it began to seem in the 1600s, a view that spilled over into science in the late 1700s, and by the late 1800s people began to think that science would soon lay out all the secrets of the universe in familiar little neat and tidy packages. A similar attitude seems to motivate many evolutionists today, although philosophers, logicians, mathematicians, and scientists in the 20th century discovered things that showed that reality is stranger than we imagined, and there are insurmountable limitations to human reasoning, logic, mathematics, and ability to understand the universe.

Back in the 1600s, however, some people began to be very impressed with the power of human reasoning. When applied to religion and science (especially geology), this faith in human reasoning produced a new view of science which, by force of definition of terms, would leave some form of evolutionism as the only possible "scientific" position, regardless of any contrary data. It was only a matter of time before someone like Darwin provided a version both impressive and yet vague enough to be widely accepted.

It all began almost two centuries before Darwin was born, and less than two decades after the first publication (in 1611) of the still-popular translation of the Bible that had been authorized by King James. In 1624, Lord Herbert of Cherbury published *De Veritate*, and kicked off the religious movement known as Deism. Deism appears to have had the aim of "saving religion" from itself. While the Reformers such as Luther merely rejected a number of traditions and practices that had grown up within Christianity, the Deists soon rejected everything about Christianity except belief in God. They kept the Bible and Jesus of Nazareth, but only as a deeply flawed human book and a good but otherwise ordinary man. What was the basis for this? Certainly no positive proofs or demonstrations, but heaps of "reasonable" doubts and criticisms.

Likewise, the first great philosophical example of unleashing reason from all constraint was not based on positive scientific information but on doubts – in this case, doubts about everything. In 1644 Rene Descartes published his *Principia Philosophicae*. He had attempted to produce a system of logic that would be free from any error that might come from misconceptions, illusions, and the failures of fallible human authorities. He found that everything could be doubted except his own existence, if only as something that was doubting. His statement "*Cogito, ergo sum*" – "I think, therefore I am" – merely puts a positive spin on this sad state of affairs. Still, from this he did build his system, which included a "proof" for the existence of God, also based on pure reason. Descartes seems to have always regarded himself as a faithful Roman Catholic, but Rome banned his books. From the other side, many Rationalists have doubted the sincerity of his professed belief in God.

At any rate, the period variously known as "the Enlightenment," or "the Age of Reason" and "the Modern Era" is often said to have begun about 1650, apparently a round figure related to the publication of Descartes' work. About this time, a growing number of controversial works published anonymously or under pseudonyms began to appear. In 1655, *Men Before Adam* argued that Adam was merely the first of the line that led to the Hebrews, and that Moses did not write the first five books of the Bible. It also proposed that the Flood had been confined to Palestine. It would be many years before such books would be published with the author's real name and become popular, but the early ones show that some people's minds were already leaping to conclusions contrary to major Christian doctrines.

The situation was only made worse by a number of defenders of Christianity who likewise set the contemporary body of knowledge and human reason above the statements of the very Scriptures they sought to defend. In 1662, a book by Bishop Edward Stillingfleet sought to show the "reasonableness" of Scripture. He argued for a recent creation of the world in six literal days, and a great Flood less than 2,000 years afterwards, but couldn't bring himself to support a global Flood. Please note that this was before any extensive work in geology. Nicholaus Steno is generally credited with beginning modern geology in 1669, and he did so within a Biblical framework that included a global Flood.

Sometimes the Enlightenment is said to have started in 1700, perhaps just because it is a rounder figure than 1650. It is also just a few years after Newton published his *Principia* (1687) and the same year that the Berlin Academy of Science began. It should be noted, in contrast, that the modern agnostic and atheistic philosophies don't seem to have affected societies greatly until well after 1700, although atheism and religious liberalism were on the rise throughout the period. It was only in the latter half of the century that some areas of scientific study began to openly depart from the record in the Bible.

There were a number of factors at work right on through the 1700s that would have provided for a generally modern world without the modern philosophies. Science was progressing on all fronts in the hands of men who regarded science as a help to traditional Christian faith. Likewise, religious leaders like Cotton Mather were also leaders of education and contributors to scientific knowledge. In several countries, governments and religious groups were becoming more tolerant of dissenting religious groups, and the persecutions and executions for heresy and witchcraft were gradually becoming things of the past. The late 1600s and early 1700s saw the production of great Christian literature such as *Paradise Lost* and *Pilgrim's Progress*, and the Christian-themed musical masterpieces of Bach and Handel. The religious revivals in England in the 1700s strongly contributed to the anti-slavery movement.

Those who opposed the Bible, Christianity, religion, and belief in God took advantage of the success and progress of the essentially Christian societies, even as they raised a great cry against all their past and contemporary failings. The greater freedom for differences in religious belief seems to have also promoted greater toleration of disbelief as well. The progress made in science and technology stoked the flames of pride in human reason. Increased knowledge of other cultures incited doubts about Christian beliefs.

In France, a period of moderate religious toleration had been replaced with an intolerance that was all the more shocking in contrast with the comparative freedom in England, and the problem was compounded by the profligate behavior of the French aristocracy. The appalling treatment of many Protestants, and of a few atheists or irreverent defilers of religious symbols, gave the *philosophes* (people who considered themselves modern philosophers) a fresh supply of powerful ammunition. Meanwhile, in spite of a number of threats, the government seems to have hardly laid a finger on the philosophes. As the government followed disastrous economic policies, especially oppressive taxation, the common people were increasingly attracted to the philosophes' promises of a new and better society. France experienced a growth in agnosticism and atheism that reached deeply into the religious leadership.

Still, the "freethinkers" and followers of such philosophers as Hobbes and Spinoza were relatively few, mostly young, and far from positions of power until the latter half of the 1700s. Pointing to the damage to society done by excesses of sin and intolerance, they blamed the very religion which, if it had been followed faithfully, would have eliminated such things. Taking advantage of the increased knowledge gathered by faithful Christians, they touted human knowledge and reason as the only cure for society's ills. From the tales of other cultures brought back by explorers and missionaries, they concocted glowing images of the superiority of cultures supposedly free from religion. Some of the works that were against the Bible, traditional Christianity, or religion in general were written by deists or formerly religious men who had been consumed by their own failures, hypocrisy, and doubts.

Few if any of these works were re-printed much after 1800, unlike Christian works such as *Pilgrim's Progress*, which continued to be popular well into the 20th century and is still readily available. Much of what was written was quickly refuted or shown to be highly questionable. The idealistic image of the peaceful, religion-free "natural" human society, for example, turned out to be the distortions of tales told for dramatic effect rather than accuracy. At one point it was "known" that writing hadn't been invented at the time of Moses, but that too is now known to be false. Some of the scoffers laughed at the Bible for featuring an entire nation, the Hittites, for which history had left no record. Archaeological discoveries since then have revealed the traces of the people in question. A few books and individual old chestnuts from this period may yet circulate in freethinking circles, but they have little effect beyond those who so strongly want to believe they don't have to believe in anything.

Furthermore, the movement started to rot from within, even as it was generally gaining in strength. Jean-Jacques Rousseau and his followers, while continuing to reject Christianity, also rejected Rationalism and other aspects of modernism, seeking instead the simplicity and emotional warmth of idealized "natural" societies and would-be Utopias. Voltaire, perhaps the greatest leader of the Enlightenment and a bitter enemy of organized religion, never strayed far from a general belief in God and a condescending acknowledgment of the practical use of religion in keeping the "lower class" content and well-behaved. The excessive claims of outright atheists repelled him. Toward the end of his life, his outward acts of religious practice increased, hypocritical as they may have been. Denis Diderot started with the disillusionment of seeing the immorality that had sprung up beneath the surface of nominally Christian Europe, and enthusiastically practiced and advocated the new "free" way of living, but strongly advised against it when his own daughter was coming of age. (For these and other insights to the period, I am largely indebted to *The Age of Voltaire (The Story of Civilization: Part IX)*, "A History of Civilization in Western Europe from 1715 to 1756, with Special Emphasis on the Conflict between Religion and Philosophy" by Will and Ariel Durant, Simon and Schuster, New York, 1965.)

In England, the reaction against Puritan rule, followed by extravagance and profligacy in the ruling class that came into power afterwards, had produced a profoundly hypocritical religious establishment that allowed gross immorality to spread through all levels of society. However, religious revival under George Whitefield and the Wesleys did much to counter the agnostic and atheistic philosophies. The religious tolerance there also gave the anti-religion movement fewer and less dramatic contemporary examples of persecution to rail against. So in England, the Enlightenment remained an undercurrent of society, although growing in influence. Unfortunately, the attempts by philosophically-minded Christians of the period to defend Christianity were so misguided or purposely deistic that many believers turned their backs on using such reasoned arguments and emphasized mysticism, feelings, and emotions instead.

Section 2: Darwinism before Darwin

It was in this volatile mix of wild new ideas, with turbulent changes (and calls for more change) in society, that modern ideas of evolution first came out, over 100 years before Darwin presented his ideas to the world in 1859.

As early as 1714, Bernard Mandeville's *The Fable of the Bees* argued that the welfare of nations depended on vices, not virtues. As support, he appealed to the idea that Nature knew nothing of good or evil, but only success or failure in a vast and fierce struggle for survival. He claimed that humans had developed language and other characteristics through this struggle for survival.

Also foreshadowing evolutionary thought, Jonathan Swift's *Gulliver's Travels* (1726) included a land where horses are the animals which developed speech and civilization, while the humans or "yahoos" are wild, filthy beasts.

In 1738, Benoit de Maillet included his evolutionary ideas in a fictional work that thinly disguised his authorship, *Telliamed*. Although he didn't publish it at the time, it began circulating among people who were open to the ideas.

In 1746 Vauvenargues' *Introduction a la connoissance de l'esprit humaine* presented ideas that would later be key components of Darwin's theory, describing "the brutal struggle for power" as "the most general law, the most immutable, and the most important in nature," in which "Nature has made nothing equal." Darwin's theory would be that the difficulties of survival, acting upon the inequalities in living things, would act like intelligent selection to produce increasingly different forms.

The year 1748 was a big one for evolutionary ideas. Benoit de Maillet's *Telliamed* was finally published, ten years after his death. In it, he proposed that land organisms had evolved from equivalent sea animals – lions from sea lions, men from mermen, and so forth. He also proposed a theory of the formation of the Solar System that was very much like the modern theory that the Sun and planets condensed out of a cloud of cosmic dust and gas. Furthermore, he calculated an age of the Earth that was closer to the current estimates than any other would be for almost two centuries. Just as many evolutionists do today, he justified his ideas as compatible with the Bible on the grounds that the relevant Scriptures were meant to be understood metaphorically.

It was also in 1748 that La Mettrie's *L'Homme machine* and *L'Homme plante* were published. These also contained a theory of evolution, based on the notion of the Great Chain of Being.

Less directly related but still important for evolutionary ideas, Catholic priest John Needham performed experiments regarding spontaneous generation and discovered microorganisms living in mutton gravy that had been boiled and sealed. Needham's experiments seemed to support spontaneous generation, in contradiction to Redi's experiments in 1668 (with meat, that showed it did not spontaneously generate maggots). In 1765 Spallanzani would repeat Needham's experiments and discover that longer boiling kills all the microorganisms, so none appear after the nutrients are sealed. Still, the microorganism version of spontaneous generation would retain a number of supporters until Pasteur performed many different experiments disproving it, in the late 1800s. By that time, however, the belief that life could readily form from non-living matter and then evolve indefinitely was already established.

Practically on the heels of the evolutionary works of 1748 came the more famous work of Buffon in 1749, his *Theorie de la terre*. He also followed a path of harmonizing his work with Scriptures, in proposing that the days described in Genesis were metaphors for ages. His estimate for the age of the Earth was only in the tens of thousands of years, but it was still far greater than a straightforward reading of Genesis allows. Furthermore, he argued against the concept of "fixity of species." Changes within and of species is compatible with Biblical creationism, but there are limits. Unfortunately, this extreme view of fixity of species, holding that every variation of life humanly classified as a species had remained totally unchanged since the Creation, was popular at the time. This idea was easy for evolutionists to attack, and so made the idea of practically unlimited natural biological changes seem all the more acceptable in contrast.

Just a couple of years later (1751), Maupertuis' *Systeme de la nature* proposed a theory of evolution that was like a sketchy preview of Darwin's work and lumped men and apes together as what today would be called "sister species." Likewise, in 1754 Diderot's *Pensees sur l'interpretation de la nature* (published anonymously) also toyed with evolutionary theorizing, along with other modern ideas.

As is the case today, evolutionary ideas extended beyond the biological realm. In 1757 David Hume published *Four Dissertations*, which included a "Natural History of Religion," positing a psychological evolution of religion. This idea would also help in the rejection of belief in the Bible and make biological evolution more acceptable to modern church leaders.

Another evolutionary theory was published in 1758. Helvetius' *De l'Esprit* (*On Intelligence*) argued that humans had evolved from animals that had begun to walk upright, thus allowing the forepaws to gradually change into hands.

The "ladder of beings" idea was again featured in 1761, with the publication of Jean Baptiste Robinet's *De la Nature*. As Darwin would do almost 100 years later, Robinet pictured Nature as an almost intelligent force, constantly working to make living things more fit to survive.

Ideas that life evolved and ideas that religion had evolved seem to go hand-in-hand. In 1765 Nicolas Boulanger's (posthumous) *Antiquite devoilee* was published. In it, Boulanger argued that religion started with primitive man's fear of natural disasters, and he promulgated the idea that religion had been organized by priests and kings to keep the rest of the population subservient. This may have been a popular thing to say in a country where the citizens were oppressed by government taxation and a hypocritical, intolerant State religion, but there has never been any evidence to back up the idea.

Another example of a pre-Darwin Darwinian was a Scottish judge named James Burnett. In 1773 he published *The Origin and Progress of Language* in which he lumped man and anthropoid apes as close relatives, and argued against the biblical picture of history in favor of a progressive struggle of Man from primitive superstition to modern civilization. Again, such ideas would resonate with people struggling to partake of the blessings that progress in science and technology were bringing about, and with liberal church leaders who saw themselves as leading the next step in religious evolution.

The next year, 1774, Diderot continued to toy with evolutionary ideas in his *Elements de physiologie*, which included thoughts on the 'missing link' between apes and men.

Buffon, continuing his evolutionary studies and publications, published in 1778 *Epoques de la nature*. This was one of the earliest works of paleontology. It set forth the notion that the different fossils found in different places and layers in the Earth indicated vast periods of time, in each of which the variety of living things on Earth was very different from that in other periods.

Most conspicuously, in 1794 a sort of Darwinism was promoted by a Darwin – Charles' grandfather, Erasmus Darwin, published *Zoonomia, or the Laws of Organic Life* at the end of his own life.

Although it had nothing to do with evolution directly, the publication in 1798 of Malthus' *Essay on the Principle of Population* should be noted here, as its speculations about population growth, death and survival provided a major inspiration for Charles Darwin's ideas about the struggle for survival and the power of natural selection.

Most famously, in 1809 (the year Darwin was born) Lamarck's *Zoological Philosophy* came out. Lamarck believed in God, but in a deistic way. He tried to explain everything as the result of a "Supreme Power" having created the laws of nature and little (if anything) else. Obviously, given such a mental framework, the only possible explanation for the existence of life as we know it requires spontaneous generation of the first life, and evolution of that life into all the known forms.

Although these early ideas of evolution were not scientific, or not accepted by the scientific community, we can see that there were a number of them proposed before Darwin. They encouraged the philosophical mindset of evolutionism. As soon as philosophers and theologians began to think that human knowledge of nature was equal to (if not better than) the Bible for learning about the past, and to think that God would not interrupt the natural laws He had set in place, some form of evolutionism was inevitable. All that was lacking was a generation or two in which to train more people to think likewise, and a respectably scientific work of literature to use as a rallying point.

Chapter 8: The New Rules Are Locked Into Geology

Section 1: What went wrong?

While the basic ideas of evolution were being kicked around in the 1700s, it was in the area of geology that science first began to cast doubt on the traditional, Biblical history of the world and prepare the way for the acceptance of the evolutionary alternative. This required a new sort of science that departed from the scientific method that was described in Bacon's writings and exemplified in Newton's. The new method was based on a philosophically expanded idea of science with lowered requirements of observation, and no experimental demonstration.

Bacon's concept, that the study of nature was like studying a second book revealing the work of God, was taken to mean that by studying nature we could read it as clearly as a book and learn all about God's work in nature. Bacon had only meant that we could learn how nature operates now, while leaving the origins of nature to the revelation provided in the Bible. In the new view, there was no theoretical limit to what could be deduced about nature, past as well as present. Once it was concluded that we could deduce the complete history of everything entirely from a study of nature it was only a small step to leaving God out of the picture entirely, but it seems that was not apparent to many people at the time.

Likewise, Newton's reverence for the Scriptures (including the passages on the creation and the global Flood) and focus on observational, mathematical, and experimental science was forgotten, neglected, or ignored. All that the new philosophers cared about was that Newton showed how natural laws governed the motions of the universe. This was satisfying to those with deistic views, who claimed that God established natural laws in the beginning and afterwards let the universe go on without interference. They argued that it was most reasonable that the Lawgiver would not interrupt His own laws with miracles like a global Flood. (Please note that this is a theological or philosophical argument.) Thus, when geological discoveries seemed to indicate things not normally possible (such as fossils of ocean life on mountains), these people ignored the Biblical history (and legends of other cultures) revealing there was an abnormal event, and thought up explanations that would require only natural processes, even if it would take vast ages and processes far too slow to observe. This new form of science formed a framework of thinking in which, ultimately, all events in the universe from beginning to end must be explained as entirely natural.

[Unless otherwise noted, the next notes were taken from *The Age of Voltaire (The Story of Civilization: Part IX)*, "A History of Civilization in Western Europe from 1715 to 1756, with Special Emphasis on the Conflict between Religion and Philosophy" by Will and Ariel Durant, Simon and Schuster, New York, 1965]

Going back in time again to see how this began to play out, we find an example in 1721, Antonio Vallisnieri's treatise *Dei corpi marini che sui monti si trovano*, which argued that no temporary flood (apparently not even one brought about by God) could account for all the extended layers of rocks made from marine sediments.

As an example of how little was known and how ready some people were to accept alternative explanations, we could take Anton Moro's suggestion in 1740 that volcanic eruptions in the sea (blasting sea creatures onto land) were responsible for the marine fossils found in rocks on mountains.

Benoit de Maillet, in his evolutionary *Telliamed* (1748), took what he believed was the rate at which the oceans were lowering, extrapolated the reverse into the past until the mountains would have been covered, and derived an age for the Earth of over two million years. Another estimate he made placed the event over two billion years ago. This appears to be the earliest example of extreme uniformitarianism. Uniformitarianism is the philosophical doctrine of assuming that there has never been any major changes in observed natural processes. This allowed practitioners to say their work and results were based on observations, and so were scientific, although what they deduced from their observations had never been observed or demonstrated to be possible. This philosophical doctrine would later become the major dogma of geology, the major reason for rejecting the young-Earth creation view, and a major component of Darwinism. (*The Faces of Origins: A Historical Survey of the Underlying Assumptions from the Early Church to Postmodernism* by David Herbert, M.A., M. Div., Ed. D.; D & I Herbert Publishing, London, Ontario, 2004)

In 1749 Buffon's evolutionary *Theorie de la terre* proposed a metaphorical day interpretation of the creation account in Genesis, with an estimated age of the Earth of 85,000 years, but in 1774 he revised this downwards to 75,000 years. This is a very short time compared to the modern estimate of the age of the Earth. Compared to the traditional age of about 6,000 years, however, the break was clear.

In contrast to all this bold new theorizing, in practice it wasn't until 1762 that Georg Christian Fuchsel produced the first detailed geological map. Nearly two centuries before radioactive isotope studies were used to try to establish absolute dates, Fuchsel promoted the idea that sets of rock layers called strata represented vast periods of time.

In 1778, Whitehurst's *Inquiry into the Original State and Formation of the Earth* provided another bold attack on the traditional view of the age of the Earth and the effects of the Flood. (*The Great Turning Point,* "The Church's Catastrophic Mistake on Geology – before Darwin" by Terry Mortenson, copyright 2004 by Master books, Inc.)

It wasn't until the very end of the 1700s or the early 1800s that geology researchers became dedicated and organized enough that the field could be considered a serious science. An academic battle over geologic theories that lasted roughly from 1790 to 1820 spurred a great deal of research and more theorizing and arguing. This is known as the Neptunist-Vulcanist debate, as researchers argued whether the continents were formed mostly by sedimentation or by volcanic activity.

The course of modern geology was set in 1795, when Hutton published his *Theory of the Earth*. More than anyone before him, Hutton was determined to set geology on the path of uniformitarianism. He believed in God, but began his geologic ruminations by rejecting any consideration a relatively recent creation and a global Flood. He even rejected the possibility of extreme changes in the rates of natural processes. He took the scientific principle of observing nature, and rationalized extrapolating it indefinitely into the past. Every aspect of geology had to be explained, in his view, by natural processes operating within the range observed in modern times. After that, it was no longer a matter of whether or not the data fit the theory, but only of finding explanations within the theory that would make the data fit.

As an important side note, it was also during this period that another field of science, astronomy, started to conspicuously leave out consideration of divine creation and possible later supernatural influences. Most notably, when Laplace published his *Mecanique Celeste* in five volumes from 1799 to 1825. It had no reference to God at all, and there is a famous story that, when he was asked about this, he said he had no need for the "hypothesis" of God.

Hutton's book was not very popular, but it was very influential in important academic circles, and in 1802 John Playfair's *Illustrations of the Huttonian Theory* expressed Hutton's ideas in terms that more people could understand and accept.

It wasn't until 1807 that a major geological society was formed, and by then uniformitarianism had already become dominant. It was also already the accepted view in most, if not all, schools of higher learning that taught geology. And so we see, it wasn't that science had found that all the data clearly showed a perfectly natural history of the Earth. Although only a small part of the world's geology had been studied, key scientists accepted the philosophical belief that the history of the Earth must have been entirely natural.

With geology given over to this belief, and evolutionary ideas already in print, we might wonder why it took so long for evolution to become accepted as the favored scientific explanation for the existence of all living things. It doesn't seem likely that the religious establishments held it back. In Germany and France religion had generally either embraced liberal views, or fallen into empty formality. France apparently came close to replacing religion with a worship of Reason in the late 1700s. While England had experienced a religious revival, it was mostly confined to the lower classes. By the 1820s, the "Cambridge network" of theologians and scientific societies were preparing people in England to embrace views of both the natural world and theology that were "progressive" both in the sense of "modern" and what they considered as "having a history of progressing from primitive and inferior to advanced and superior." Having rejected the authority of the Bible, many clerics and young scientists would eagerly embrace a theory which held that their religious beliefs, their species, and their society were each the latest and greatest in a long line.

So what did hold back acceptance of evolutionary theory? Perhaps one major factor was the association of such ideas with the French Enlightenment and the French Revolution, which had so quickly proceeded to the idolization of Reason and the shocking Reign of Terror (1793-1794), in which tens of thousands of French citizens were killed, often for little else than being too rich or religious. (Ironically, at the same time Thomas Paine's *Age of Reason* was published, praising the superiority of Rationalism over religion.) However, the major factor was probably that most leading scientists, especially biologists, saw in nature more reason to believe that living things had been created directly as wonderful works of God than through gradual natural processes.

Section 2: The Science that Stayed on the Tracks.

In the next chapter we'll look at how the same speculative, philosophical, and anti-biblical views described in this chapter continue to frame the debate. First, however, I'd like to back up and present some examples of men involved in the beginnings of science who held active, traditional religious views. These men proclaimed that they were studying the handiwork of God. I'll also point out some of the great advances in science that were accomplished during this time, within the discipline of the scientific method. I am especially indebted to the book *Men of Science, Men of God* (MoSMoG) by Henry M. Morris, (copyright 1988, 13th printing 1997) for much of this information.

I'll start with Increase Mather (1639-1732), because he is often portrayed as no more than an ignorant puritanical preacher, but this is in spite of the fact that he was "an avid avocational astronomer and promoter of science" at a time when many amateur scientists were on a par with the professionals, and he was also "the primary founder of the Philosophical Society and one of the first presidents of Harvard."

Then there's Robert Boyle, author of *The Skeptical Chymist* (1661) and "one of the founders of the Royal Society of London." He "is generally credited with being the father of modern chemistry." What is much less often noted is that "he was also a humble, witnessing Christian and a diligent student of the Bible" and "profoundly interested in missions." Another of his significant works, *Concerning the Usefulness of Experimental Philosophy*, was published in 1663.

The Royal Society of London itself, which was "granted a royal charter" in 1662 by Charles II, was largely inspired by Francis Bacon's fictional account of an institution named after King Solomon and dedicated to the study of the six days' work of God in creating the universe. This was reflected in the real society at its founding. "As a group, they believed that God was the omnipotent Creator and that they, through their scientific endeavors, could reveal to the world the grandeur and the beauty of His creation." (Reference: *The Faces of Origins*)

Let's not forget Redi and his experiments in 1668 which demonstrated that maggots didn't form spontaneously from rotten meat, but hatched from eggs laid on the meat by flies.

Nicolaus Steno, who lived from the 1630s to 1680s, began the systematic study of geology long before geology was organized into a field of professional science, and he did so within a global Flood framework. He began to make these significant contributions around 1669.

In 1675 The Greenwich Observatory was established, with John Flamsteed in charge. Flamsteed (1646-1719) "was also a faithful clergyman..."

It was in the late 1600s that Isaac Newton (1642-1727) published his great works on optics, gravity, and motion. He was elected president of the Royal Society in 1703. Although he had unorthodox views on one or two religious doctrines, he was as strong a believer in the Bible as anyone. He "wrote strong papers refuting atheism and defending creation and the Bible. He believed that the worldwide Flood of the Bible accounted for most of the geological phenomena, and he believed in the literal six-day creation record" (MoSMoG). In 1728 a book that Newton wrote was published (posthumously, of course), in which he explicitly supported divine creation of the world less than 6,000 years before (*The Faces of Origins*).

Probably very few Americans would be able to tell you anything if you asked them about "John Ray, the seventeenth-century scientist," but according to *Men of Science, Men of God* (and other sources) he "has been called the father of English natural history." He was born in 1627 and died in 1705. (Note that this is over 150 years before Darwin's *Origin*). During his life he "was one of the founding members of the Royal Society. He was the greatest authority of his day in both botany and zoology." In 1691 his book, *The Wisdom of God Manifested in the Works of Creation*, proclaimed the triumph of Natural Theology. Natural Theology is the study of how the wisdom and power of God are displayed in nature, God's "second book" of revelation. Ray's work showed, for example, the "intricacies of the human body" as far as they were known at the time, and "it was to constitute the chief obstacle to the rise of evolutionary views."

Even today, many of the evolutionary attacks on creation science and Intelligent Design amount to no more than philosophical or theological arguments of dysteology, that is, pointing at what they believe is wrong with things. Besides the fact that these arguments are as non-scientific as the Natural Theology they attack, they all have one or two basic weaknesses. In the first place, they make assumptions that body parts could be better rather than demonstrating superior alternatives. (Nobody has built an artificial replacement that people want instead of a healthy, complex natural part. There's no accounting for taste when it comes to cosmetic replacements and alterations.) They often ignore the fact that a good design may have to account for many needs, functions, and potential problems. Even Intelligent Design proponents often ignore the disastrous changes to the original creation that came about through Mankind's sinning, rejecting the Creator and His special help in maintaining the creation. The degeneration of our bodies, the destruction of the original biosphere by the Flood, and an untold amount of cruelty and inefficiency in nature would never have come about if it weren't for sin. Criticisms of the way things are also overlook our inability to comprehend the plans, purposes, and designs of such an awesome Creator.

Recognizing the limits of science and the effects of sin was not a problem at the end of the 1600s. Dr. John Woodward's *Essay Toward a Natural History of the Earth*, published in 1695 (100 years before Hutton's major work), was "based on a complete confidence in the trustworthiness and historicity of the Bible." Dr. Woodward saw that the global Flood "accounted for the stratification of the earth." He also derived from the Bible "the most plausible of scientific hypotheses concerning marine fossils, differing species of trees and the American people." (*The Faces of Origins*)

Cotton Mather (1662-1727) was the son of Increase Mather and, like his father, is likely to be portrayed as an ignorant religious fanatic. In truth, he followed in his father's footsteps in promoting education and engaging in scientific observations. He was "president of Harvard" and "probably the first American to publish original contributions in science, with many publications in the Transactions of the Royal

Society." (*Men of Science, Men of God*) For example, in 1716 he "reported from Boston to the Royal Society of London a demonstration of hybridization by wind pollination." (*The Age of Voltaire*)

Another of the great scientists of the time who, like John Ray, is now unknown to most Americans, is the physiologist Albrecht von Haller. Before his great works in physiology, he "issued a volume of lyrics, *Die Alpen*," in 1729. This "book anticipated Rousseau in almost everything. It invited the world to admire the Alps both for their inspiring elevation and as a testimonial to God; it denounced cities as dens of luxury and irreligion ... it lauded the peasants and mountaineers for their hardy frames, sturdy faith, and frugal ways." It was "in his later years" that he "placed himself at the head of his kind" (physiologists) "by his *Elementa physiologiae Corporis humani*, which appeared in eight volumes between 1757 and 1766." It was "one of the century's scientific masterpieces." (*The Age of Voltaire*)

The eighteenth century doesn't seem to have produced many scientists of the caliber of the greatest ones in preceding and later centuries. One of the few was Karl Linne, or Carolus Linnaeus (1707-1778). In 1736, "Inspired and helped by [Boerhaave,] that nobleman of science, Linnaeus issued one of the classics of botany, *Systema Naturae*." (*The Age of Voltaire*) This began the system of classifying all of life on Earth which is used (with some adjustments) today. According to *Men of Science, Men of God*, "One of his main goals in systematizing the tremendous varieties of living creatures was to attempt to delineate the original Genesis 'kinds'."

As a first attempt, it was a great work, but it set the requirements for species so low that there is difficulty distinguishing a species from a mere variety. Separate species have often interbred successfully when given the opportunity. While Linnaeus himself recognized that his "species" might not represent the limits of variation of the original created types, evolutionists would still come to regard any departure from a strict fixity of species as evidence for their belief. Many evolutionists today consider variation, even within a species, to be strong evidence that microbes could be the ancestors of Mankind.

Here's another example of the work of a "religious man" who also demonstrated a comprehension of science while giving honor to God: "The Abbe Pluche spread out the *Spectacle de la nature* in eight volumes (1739-46); it ... displayed the wonders of science, and the evidences of design in nature, to manifest the existence of a Deity supreme in intelligence and power." (*The Age of Voltaire*)

If the 1700s were a bit thin as far as great scientists go, the latter half likewise seems a bit thin on scientific advances. The *philosophes* and their so-called Enlightenment came to power in France and spread their ideas abroad, but did not include any great scientists in their number. Joseph Priestley, an English chemist who embraced a number of the same ideas, did make several important discoveries in chemistry. On the other hand, the French chemist Lavoisier was one of the victims of the Reign of Terror. Meanwhile, some of the British colonies in North America were busy with a revolution of their own, while Britain itself was busy gearing up for the full flowering of the Industrial Revolution.

This period did see a number of significant applications of science, such as the lightning rod, the development of steam engines, and hot-air balloons that could carry people. It seems, however, that the philosophy of the age had already invaded several areas of science and, with some great and notable exceptions, more scientists were accepting the new idea that science was the effort to explain everything naturally, not just to study nature as it is currently operating. In most areas of science, there's no practical difference, and so great strides continued to be made, once the social turmoil settled down. Several of the great scientists at this time who continued to stand for the original divine creation, had already rejected the global Flood and relatively recent creation, due to their faith in human ability to "read the record" in the rocks. But by the end of the 1800s, there were fewer leading scientists who openly espoused Creation.

Throughout this period, however, there were some exceptions worth noting. In 1768 Catcott's *Treatise on the Deluge* presented a geological defense of a recent creation and global Flood. (Ref.: *The Great Turning Point*) In 1781 William Herschel (1738-1822) discovered a new planet – something that had never been done before. According to Morris, Herschel "has long been recognized as both an outstanding Christian and an outstanding astronomer. In astronomy he made many great discoveries, perhaps the most notable being the recognition of double stars and the discovery of Uranus. ... As a Christian, Sir William was...noted for his kindness and his sublime conception of the universe as a marvelous witness to the handiwork of God. It was Herschel who said: 'The undevout astronomer must be mad'."

One of the last great geologists to work within the biblical framework was Jedidiah Morse (1769-1832). According to Morris, he was "the leading geographer of America during his lifetime. He wrote the first American textbook of geography, almost universally used in the schools of the day and going through 25 editions, many of them after his death. He was a strong advocate of flood geology and the literal-day Mosaic chronology of earth history." In case you are wondering about the familiar last name, he was the father of the inventor of the telegraph, Samuel Morse.

In 1796 Georges Cuvier (1769-1832) began the science of comparative zoology. The similarities between various organisms is now usually claimed as an evidence that they evolved from a common ancestor, but Cuvier recognized that there were also great differences, and that similar patterns, themes, and motifs are often found in the works of artists and other creators. He argued against the evolutionary ideas of his time. However, he accepted the belief that the vast geological formations must have been formed over a vast period of time. He held that each set was formed relatively rapidly and didn't deny the global Flood of the Bible, but he promoted the idea that there had been many such catastrophes spread out over vast ages.[*The Timetables of History*, MoSMoG]

There were two important works on each side of the turn of the century which probably helped put off acceptance of evolutionism. *Scriptural History of the Earth and Mankind* by Philip Howard "defended the biblical account of a global deluge" in 1797, and in 1802 William Paley's *Natural Theology* argued that "There cannot be design without a designer" (without any references to the Bible). Paley's work was the first to make the argument that life was designed by the illustration of finding a watch far from any human that might have made it. Even if you didn't see it being made and never knew who made it, and never saw another watch that looked like it, you would still conclude that it was made and not a natural object. Things like watches just don't form without the application of intelligence. The basic idea is still around because, although evolutionists devise clever stories about how such thing might have evolved, they have never demonstrated exactly how. Meanwhile, we keep discovering new amazing features of living things and whole new levels of complexity.

Chapter conclusion: The stage is set for Darwin

After the early 1800s, while there were still great creationist scientists like Michael Faraday and Louis Pasteur, the dominance of the new philosophy of science in the universities and research societies meant that it would become almost impossible to study, practice and publish scientific research that went against a thoroughly natural history of the universe. It didn't appear to be atheistic, thanks to the deistic phrases commonly used. It certainly wasn't atheistic on purpose, for then atheists were a small minority of the population, among scientists as well as the general population.

Progress was the spirit of the age, and nothing showed progress more than science, and science had come to be thought of as the attempt to give natural explanations for everything. The only perfectly natural explanation for the formation of the Earth has to be some sort of nebular hypothesis and uniformitarian geology. The only perfectly natural explanation for living things must be some form of gradual evolution, starting with raw chemicals. If everything could be explained as perfectly natural, Laplace would be right: what need would there for the hypothesis of God?

Still, while the Deists and liberal Christians accepted the naturalistic explanations for the origin of Earth and its geology, and some form of evolution, they didn't like the notion that God had nothing at all to do with the origin and diversity of life. They joined with the conservative or traditional believers in drawing the line there. Even doubters of the Bible and "freethinkers" such as Voltaire found the atheistic origin of life too much to believe.

Nor was believing in a purely natural origin and history of life needed for biology. On the contrary, a few men as familiar with geology as anyone of the time continued to advocate the biblical Flood as the explanation for major geological features well into the 1800s. However, their views were almost entirely shut out of the universities, geological societies and journals.

In 1760 Voltaire had written to Helvetius, "This century begins to see the triumph of reason." Less than 60 years later, naturalistic Rationalism was at the heart of most advanced education in geology in Britain. Given the progressivism and scientism of the age, acceptance of evolutionism was inevitable. It was just a matter of time and coming up with clever ways to explain away, minimize, or obscure the problems and missing evidence. Given the number of clergymen who were also ready to throw out a "literal" reading of Genesis (and the rest of the Bible) in order to appear modern and progressive, it's no wonder that by 1860 a theory of evolution had been published that quickly gained acceptance in the proper circles. That story will have to wait for a later chapter.

Chapter 9: Rules have consequences

Introduction: Keep the Problem in Proper Perspective

As I've shown so far, over a period of about 200 years (from the early 1600s to the early 1800s) science as it is generally thought of today was formed in two steps. First, "natural philosophy" and individual studies of nature turned into "science," an organized process for studying the way nature works that was about as independent of philosophy as it could get. (Any good philosopher should be able to explain to you how pervasive philosophy is.) It stuck to things that could be known by repeated independent observations and experimental demonstrations. The second step was the re-introduction of philosophy through a back door. In the guise of merely extending the practice of science (studying nature), some scientists introduced the philosophical assumption of naturalism. This was heavily influenced by both the theological doctrine of Deism and the philosophy of Rationalism.

For most scientific studies, this change was totally irrelevant. The great consequences it has had apply to a few new areas of study that were added to science, and to general ideas and attitudes about science, philosophy, and theology. Along with the dependence on the philosophy of Naturalism, as science revealed more about nature, came a greater faith in the pronouncements of scientists. For many people, this seems to have amounted to believing that scientists might one day find the answers to all the major questions about everything, past and future as well as in the present.

No one knows how many have consciously held this attitude (that has been called Scientism), but it probably has affected everyone in modern cultures to some degree. Many people with very strong religious beliefs wouldn't dare profess any position which contradicts scientific authorities. It's a strong and widespread belief that "so many scientists can't be wrong," although the majority have been wrong about things sometimes.

Two modern advances supporting belief in billions of years

There are two powerful beliefs which cannot be tested according to the scientific method but are now widely held to be undeniable scientific facts. They were developed after science first began to expand beyond its proper limits, and included rationalizations that assumed Naturalism. Both beliefs require the assumption that what we observe can reliably tell us about a past long before human records. These two beliefs are not directly related to biological evolution, but, like the uniformitarian geology that came before them, they make unimaginably vast stretches of time a "scientific fact," and to be at all believable, any naturalistic theory of evolution needs all that time for life to evolve. Billions of years are also not indicated by a straightforward reading of the Bible. However, some creationists do make room for them.

The first is the assumed relationship between starlight and the age of the universe. The other is the assumed reliability of the relationship between various radioisotope ratios and the age of the materials in which they occur.

Theologically, and in general philosophical terms, there's no real reason to find these beliefs undeniable, but when confined to naturalistic philosophy, it's virtually impossible to deny them. If nature has always operated as it does now, then an age estimate for the universe, and of Earth, of billions of years is unavoidable. However, it's presumptuous to assume that the laws of nature have always been exactly the same, or that supernatural forces have never altered the course of nature.

When science began, nobody knew how far away stars could be, nor did they know the speed at which light travels. For all that anyone could tell at the time, light appeared the instant a source was created. As physicists measured the speed of light, and astronomers began to realize just how far away stars could be (this didn't happen until the 20th century), there was a deduction to be made that the "scientific-minded" consider inescapable: the universe must be very old, because it would take that long for the light from those very distant stars to travel to us. There are several ways to explain how we can see stars so far away if the Earth is only thousands of years old, but for those who insist that scientific facts can be discovered by assuming the universe began more or less naturally, and current natural laws have remained constant for the vast majority of the history of the universe, the case is closed. Such an insistence is a personal, philosophical stance, but anyone who wants the comfort of an unassailable position (AKA a closed mind), can find support from the major scientific establishments that this is the proper scientific view.

The other strong argument for the universe being old enough to make Evolution seem plausible is radioisotope-based dating. In the twentieth century, scientists discovered that certain elements were falling apart at a detectable rate. As they shot out subatomic particles and radiation, they changed into varieties of other elements that were also unstable and continued to fall apart, until at last they reached a stable form. For example, uranium eventually breaks down into lead. It seems obvious that we should be able to tell the age of a rock that has such elements in it by measuring how much of each kind there is. In practice, since many steps don't take very long, only two elements or different forms of one element are measured. This method always shows rocks to be at least millions of years old, and the oldest dates calculated are several billion years. On the face of it, this method must give dates as reliable as the measurements of the speed of light.

Vast ages is the first natural deduction from observing nature

But before these two indicators of great age, there were a number of others which, while not having the impact granted by the reliability and precision of physics, have been very persuasive. These are the geological data that, under uniformitarianism, were the first to cast doubts on the traditional belief that the Earth was about 6,000 years old.

First there's the sheer volume of sedimentary rock. It's not hard to look at the slow formation of river deltas, lakes gradually filling up with sediments, and similar processes to get the idea that they would take vast ages to form all the sandstones, mudstones, and other sedimentary rocks. There are some areas with vast deposits of volcanic rock, too. Some large rock formations are made of materials that we commonly see forming deposits as the water they were dissolved in evaporates. For example, salt deposits thick enough to mine using large trucks. Clearly, if these formed by the ordinary process of cycles of evaporation, it would have taken a very long time.

Then there's the observation that the rocks in the earth have many layers (strata) in them. Some rocks have many tiny layers that are generally understood as the work of one year or at least six months each, and in some places, larger layers alternate between river and ocean sediments. Ordinary processes would take a very long time to form similar structures.

Many people find the general pattern of the fossils to be persuasive, not only of vast ages but biological evolution. The rocks calculated to be the oldest have no fossils, or only traces of bacteria, such as stromatolites: piles of mats of sand grains stuck together with organic material. Rocks assigned younger dates have more complex organisms. Living kinds of mammals and birds only appear in the most recent (Cenozoic) group of layers. It's easy to see in this pattern, in the mind's eye, the story of life evolving over vast ages of time.

Another type of evidence which is said to be possible only if the Earth is very old is the sheer volume of organic matter indicated by fossils and other materials in the earth, such as coal, oil, and natural gas. Calculations indicate that it is far more than exists in the world now. In the standard Young-Earth creation framework, most of the organic material trapped in the rocks would be from what was alive at the time of the Flood.

One of the first things to cause some people to doubt that the Earth is only a few thousand years old came from legends and translated histories from non-European cultures and ancient texts. These other cultures' histories and legends were longer than genealogies in the Bible. Later, carbon isotope dating and other methods also indicated that early civilizations or settlements were somewhat older than would be possible, given a relatively recent global Flood, or even a date for Creation of about 6000 years.

Added together, these arguments or evidences seem to overwhelmingly prove that the Earth is unimaginably older than a few thousand years, and the universe as a whole must be several times older still. From this view, it's no wonder that evolutionists and old-Earth creationists can't understand how anyone could hold to the young-Earth position.

Choosing a foundation to stand on

So how can the YEC position be held without rejecting science in general? I start by believing in God, the eternal and infinite basis of the existence of everything else, and the creator of the universe. Science cannot prove or disprove the existence of such a being, but I believe that what we've learned through the scientific method about the operations of nature show that such a Being is the best explanation for the existence of the universe, and the only possible explanation for the existence of living things and all their forms. Then it is logical to suppose that such a Creator, in creating humans, would provide information about himself and his work of creation, beyond these basic points which show his power, intelligence, and authority. I have found no better candidate for a record of this information than the Bible. I have faith that God would communicate to us clearly from the very beginning, just as others have faith that nature has "always" operated consistently, without a supernatural origin or supernatural intervention. I don't believe God would tell those who believe in him something that would leave them with a false impression of what he did, for most of the thousands of years that we've had records of it. In other words, I trust God to have communicated with us directly in a way that is more clear and reliable than our ability to puzzle out the facts of the past by examining the evidence.

The existence (if it can be called that) of billions of years of the past cannot be demonstrated by the scientific method of repeated observation and experimentation. What seems obvious and undeniable depends on certain assumptions. Once you recognize the uncertainty of these assumptions, you can understand that however much data we gather, and how consistent it all seems, we still can never be sure the assumptions and rigid, narrow thinking haven't led us to false conclusions.

A Helpful Illustration, with my own twist and development

Some questions simply can't be truly settled by science, and setting a date for the beginning of things that happened long before recorded history is one of them. If that last sentence is hard to swallow, perhaps sounding like the scientific equivalent of heresy, I could easily give you a demonstration to illustrate its truth. Other creationists have used this two-candles illustration, but I have found a way to carry it out that is a little more elaborate and thorough, and makes a point or two in addition to the problems of the unknowns of a singular past event.

Put two identical candles in a bare room. You pick the candles, making sure they are identical. Make sure the room is empty. Search me to be sure I have nothing hidden that might be used to alter the length or rate of burning of one candle. Give me a match. Allow me to be in the room alone for awhile, at least long enough for a candle to burn down to a certain noticeable degree, such as a quarter or a half its length. Check me again when I leave. Check the room when I leave and there will only be some melted wax, the burned match, and the candles. The candles will be different lengths, easily visible without measuring. Now, no scientific investigation you can do will correctly answer the question, "Which candle burned for a longer time than the other?"

You might think it obvious that the shorter candle burned for a longer time. You might list a number of reasons for thinking this must be true: you know the candles were of the same length to start with, you know that they were composed of the same sort of wax and wicks. You checked me before I went in so you know I didn't make one candle burn faster with oxygen or some other gas or liquid. I didn't have a knife or anything to cut one candle and didn't have any part of a candle with me when I left. There was no cut or broken-off part of a candle left in the room.

Now suppose you ask other people the question. You can show them the candles, and if they have questions, share all these facts – except your knowledge that I had some reason for doing something special with the candles. Wouldn't they all answer "The shorter candle?" Even knowing I had a point to make, you still might think I was only bluffing, and the shorter candle must have burned for a longer time than the other. It might seem crazy to say it hadn't.

Of course, you'd not only be wrong, you'd be trying to answer the wrong question. Although you knew I might do something tricky to make a point, you would never come up with the right answer if you only tried to figure out which candle burned longer. You'd also have to consider that the question was misleading you with a hidden assumption.

You may be thinking, "Yes, we know you may have done something clever with the candles, but science only studies the predictable forces of nature." This is exactly my point: certain areas of science have been infected with the **assumption** that **nothing** intelligent **ever intervened** in the processes of nature. Ever. But remember, this is a relatively recent philosophy. Less than 300 years ago, almost everyone in Western civilization believed that God created the heavens and the Earth in a special way, not by natural forces that we can measure and calculate.

You might say, "But even if God created things supernaturally, He wouldn't play any tricks on us!" However, I might have a reason for wanting candles of different lengths and use the same technique, without any thought of making a point about estimating time or trying to fool anybody. The assumption that God created in such a way that we can figure out the entire history of the universe is what led to naturalistic thinking in the first place. It ignores the fact that instantaneous supernatural creation of a complete universe is not possible by natural processes and would inevitably include things that couldn't be explained naturally apart from vast periods of time, if at all. The naturalistic philosophy that entered science in the late 1700s is a willful ignoring of that possibility. I must emphasize: this did not begin with scientific demonstrations, but was based on a theological argument, presuming that we can know what God would (and wouldn't) do in creating and sustaining the universe. Consider how ordinary things like instant cameras and even mirrors might seem strange, magical and deceptive to people who had never seen them before. Things that God could do supernaturally in the most simple and direct way might seem strange and possibly deceptive to the most intelligent humans if they limited their analysis to natural processes or merely human rationalizations of what God should or shouldn't do.

Some may argue that we have consistent, independent lines of evidence that eliminate the need for philosophical or religious considerations. This argument would only be persuasive if we could know we had considered all possibilities, and if the consistency and reliability of the evidence was practically perfect.

Consider again how my simple illustration would give a candle-dater many reasons for confidence. Now let's see if there are any indications that uniformitarian stories have significant weaknesses, and that the established dating methods are not absolute. Could it be that most of the scientific establishment is willfully ignoring or weakly excusing away problems? Remember how in physics and astronomy, things seemed to be nearly settled and complete in the late 1800s, but a few exceptions, unanswered questions, and a puzzle or two led to great changes in how we viewed the world around us.

Looking at things from a different perspective

While we can't repeat the billions of years claimed to have gone by, we can examine the evidence from outside the box of naturalistic assumptions and expectations. There are a number of considerations which are more in harmony with the Young-Earth Creation than with billions of years of unvarying natural forces. Some of the arguments used against YEC beliefs may be in reality more easily reconciled to them than to the idea that humans evolved from animals over vast periods of time.

For example, some legendary histories of ancient cultures may be longer than the Biblical chronology, but are still much closer to YEC time than the hundreds of thousands of years of evolutionary human history. Plato set his mythical Atlantis less than 10,000 B.C. One of the popular suggestions for a real civilization that inspired the legend was destroyed by the eruption of Santorini less than 4,000 years ago. To accept the dates given by some legendary histories, we would also have to accept the very long lifespans for ancient individuals, which naturalistic science rejects. If you reject the Bible's accounts of people living for hundreds of years, you can't consistently oppose its chronology with genealogies or king lists that wouldn't be so long without those extremely long lifespans. The ancient calendars of China and the Incas (and of course the Hebrew calendar) are also consistent with the YEC view.

Estimates for the domestication of various animals also fall close to the YEC framework, although based on old-Earth assumptions and dating techniques. The oldest estimation for the domestication of wolves as the first step toward modern dog breeds, is only about 15,000 years ago. Most of the other cases are supposed to be under 10,000 years, with most falling within the historical period of 4,000 years. Most breeds of dogs began after 1800, and hamsters weren't taken in as pets until the 1930s. Remember, some fossils of our species have been dated about 300,000 years old. Farming, too, is dated to a relatively recent 11,000 years ago, although evidence that humans were harvesting and using wild grains for food is dated more than 11,000 years earlier than that.

Human population growth studies are also compatible with a young-Earth creation framework. Even evolutionists have to admit that it is quite possible for the human population to grow from a few individuals to its current size within a few thousand years. It's true that the population could grow much more slowly or experience more crashes, and thus work within an old-Earth framework, too, but that seems unlikely on several grounds. Most of the possible causes of population limitation or reduction would probably wipe out a small population. Once a population became large and successful enough to be safe from extinction, it would most likely spread to its full potential quickly, and be less susceptible to crashes and other problems limiting its growth. Our ancestors showed a lot of potential hundreds of thousands of years ago, according to the uniformitarian-materialist story.

Many people think that carbon (C14) dating (comparing the ratios of carbon isotopes in organic materials buried in the earth) is an absolute and reliable technique. It should be obvious, however, that if there were much less radioactive C14 to start with, then this technique would have yielded large "ages" from the beginning. A number of cases have been documented in which living or recently dead organisms had very low amounts of C14, and therefore great apparent ages. Likewise, evolutionists wouldn't even consider checking for it in fossils found in layers believed to be too old. Creationists have tested samples that evolutionists would consider too old to use this technique on, and gotten definite results – fossils supposedly millions of years old getting dates with a technique limited to tens of thousands. Yes, it's "not supposed to work that way," but that's the point – if you get results with a measuring technique that's not supposed to work, then there's something wrong about the technique or your assumptions about it.

Counting tree rings (traces of growth and dormancy through changing seasons), seems like a reliable method. However, this also depends on the uniformitarian assumption. If the climate in the past went through a period of turbulence, trees could be forced into additional, short periods of dormancy and growth, creating more rings and the appearance of more years. If these changes happened on a global scale (or where tree rings have been counted), then all the trees would have similar patterns. We've never observed such conditions, so we wouldn't know how to recognize the effects. Living trees have at most a few thousand rings. Older tree-ring dates are based on comparing such rings to those of dead and buried trees. This is a subjective process of pattern-matching. It is well-known that humans have a great propensity to find patterns, and similarities between patterns, even where any similarity has no real significance.

Similar to the tree-ring technique is the attempt to date the age of ice by studying tiny layers in glaciers and ice sheets. Like counting tree rings (only worse), this method involves a subjective examination of tiny traces and judging them by human pattern-matching and old-Earth assumptions. Isn't it strange that something as variable as the precipitation, accumulation, and preservation of something as ephemeral as snow is considered a reliable indicator of age? After all, in the uniformitarian scheme, there have been many Ice Ages and warm spells, glacial advances and retreats, not to mention the historically recorded variations. Does it seem at all likely that we can rely on every thin layer of dust to represent one year or season? There is a well-documented case of a flight of P-38 fighter planes that were forced to crash-land in Greenland, and in a matter of decades were buried many feet deep in the snow ice. (http://p38assn.org/glaciergirl/index.htm) Artifacts in Antarctica have suffered a similar fate (https://en.wikipedia.org/wiki/Antarctic_Snow_Cruiser#Rediscovery_and_final_fate). Likewise we have recently seen much melting of glaciers and ice sheets. Could it just be in that area at that time that there was so much snow in a short time? Can it really be expected that any place on Earth has had nothing but slow, consistent accumulation of snow for many thousands of years?

Other techniques seem to have come and gone, gaining popularity as they seem to support some theory and losing it as they produce inconsistent results or threaten established opinions. Recall the case (Chapter Two, Section Two) of human or anthropoid footprints in Mexico in strata dated to the time before humans were supposed to have arrived, so it was quickly attacked by claims that the dating was off, or that the footprints were misinterpreted. A different dating method produced a much older age (and eventually it was determined that the marks weren't footprints of any kind). If the proponents of these methods can say that they are fallible, why shouldn't skeptics suggest perhaps the situation is much worse?

Even established dates, presented to the public as scientific facts, have had to be tossed aside. Dates for "prehistoric" artifacts and sites, that were generally accepted at one time, have been more recently challenged and sometimes changed, to the extent of being about double or half their earlier estimates. Some finds of "prehistoric" bones or art have turned out to be less than 2,000 years old; one or two "ancient" finds turned out to have been produced within living memory. Then there are the cases of outright fraud, and several examples of simple but large changes in dates that have been kept mainly within the scientific community. (For example, compare "Chauvet cave's art is not Aurignacian: a new examination of the archaeological evidence and dating procedures," by Jean Combier & Guy Jouve, *Quartaer* 59 (2012): 131-152. http://www.quartaer.eu/pdfs/2012/2012_combier.pdf with "A high-precision chronological model for the decorated Upper Paleolithic cave of Chauvet-Pont d'Arc, Ardèche, France" by Anita Quiles et al., PNAS vol. 113 no. 17, 4670-4675 http://www.pnas.org/content/113/17/4670.abstract)

The general response by old-Earthers to such observations is that they merely show the self-correcting nature of science, and this is supposed to reassure us that this area of science is reliable, because it corrects itself. Furthermore, sometimes the dates have been made greater. However, we're looking at the bigger picture – are scientific techniques capable of revealing the past as if we had a time machine, or can we see indications that all is not so factual and reliable as we are told? On this question, even those changes that increased the estimated ages show that the "dates" were (and the new ones still are) subjective estimates rather than objective facts. If "facts" have to be corrected, then they never were facts to begin with. Why should we trust that the latest such "facts" are truly facts now? How can we be confident the corrections are correct? If it seems wrong to incite questioning the pronouncements of scientists, maybe it's because they have become like a new priesthood, rather than fallible humans seeking tentative answers.

Even the dates of some significant geologic events are dated relatively near the YEC time frame, despite having been derived by uniformitarian assumptions. The ("most recent") Ice Age supposedly ended no more than 12,000 years ago. The region of the Sahara Desert was lush with plant life and large lakes as recently as 6,000 years ago. Forgive me for the repeated reminders, but this is judging only by the usual old-Earth age-reckoning. Beginning in the 1500s and lasting into the 1700s, there was such a widespread and dramatic cold spell it is known as "the little Ice Age." Since such large and dramatic changes have taken place within the last eye-blink of the uniformitarian time-frame, it appears to be further grounds to question the validity of the general uniformitarian assumption.

Then there are the huge and precisely-built pyramids, monuments, and other works of ancient civilizations which remain amazing, especially given the presumed "primitive, ignorant" state of humans involved. Although standard dating places them a few thousand years too old to fit in the YEC time frame, they are still quite recent compared to the supposed origin of physically modern humans. How is it that the first civilizations included movement of huge stones and works of architecture that are still impressive today?

There are also cultural artifacts, legends and some ancient works of art that suggest that ancient people encountered living dinosaurs, or recalled stories of men who had. Similarly, there are traditional legends known from all over the world that have a number of similarities to the Biblical account of a global Flood.

Another human-related indication that something is wrong with the Old-Earth view is the existence of out-of-place artifacts, or OOPArts. Some of these are dated after the rise of known civilizations, such as the mechanical astronomical computer known as the Antikythera device, and what could be an early form of battery found in the Middle East. Nobody doubts the authenticity of these finds, but they indicate knowledge of things thought to be known only much later, and their full significance is more or less controversial.

Other cases range upward in supposed age and "out-of-placeness" to include footprints and metal objects found in rocks that are commonly dated to the dinosaur age or even much earlier. Some OOPArt cases seem as elusive and controversial as UFO sightings or Bigfoot. In contrast, in several cases the evidence is solid and observable, and skeptics can only cast aspersions on the accounts of how they were found. Other examples can still be seen in place, and the only controversial aspects are matters of opinions and subjective interpretations. At any rate, it's a fascinating subject and you may be surprised by some of the cases.

But what about the things that seem to be solid evidence for vast ages? If you look with different assumptions and expectations, the general fossil pattern and geologic structures indicate the results of a global catastrophe better than a vast sequence of traces from "business as usual" natural forces. Right now I'll just point out some problems for the materialistic philosophy framework and some things that fit well with a global catastrophe.

Problems and special explanations in the old-Earth view

Perhaps the most persuasive evidence presented in support of an old Earth is radioisotope dating of rocks. This seems more straightforward and reliable than C14 dating of organic matter. If we find radioactive materials in the earth along with their decay products, it seems we should be able to calculate how old the rocks are according to their decay rates.

Strictly speaking, dating the age of the Earth is something of a matter of faith even within the naturalistic framework. Scientists discovered radioactivity relatively recently. Studies leading to modern dating methods didn't produce fruit until about the middle of the 20th century. So we've been studying radioactivity and using it for dating for only an eye-blink of time even compared to recorded history. In that time we've discovered Dark Matter and Dark Energy, and a different rate of expansion of space at the beginning than now. How can we be so sure that these techniques can reliably give us dates hundreds of millions of times greater than the time we've been using them?

Consistency of results can be good support, but is isn't so much when consistency is largely built into the philosophy and practice of all the dating techniques. Inconsistent

results are rejected without consideration, leaving a false appearance of consistency. And in development and practice,
radiometric dating techniques aren't simple matters of testing and calculation. They are calibrated and judged by comparison with fossils. Apparently reliable dating results which have not sat well with established views have been challenged and different tests have been done until acceptable ages were produced. C14 dates beyond a certain limit are automatically given an adjustment to account for some known skewing factors, but it is assumed that there aren't any others.

If an undesirable date can't be replaced by other methods, or the other methods produce dates even farther from expectations, there are a number of excuses to make such "wrong" results go away. Some isotopes might be assumed to have been partly removed by leaching – washed out by water filtering through the rock. Or special conditions might be invoked that supposedly opened tiny cracks in the minerals to allow other isotopes in, etc. There seem to be escape hatches and exemption clauses for every technique, when needed, but when the establishment is happy with the dates, then it's not likely that anybody will see if different dates might be obtained by other methods. It does sometimes happen. Even when within an acceptable range, dates have been changed after further research to about twice or half the original figure. For an extensive, technical examination of such matters, see John Woodmorappe's book, The Mythology of Modern Dating Methods (Institute for Creation Research, El Cajon, California, 1999).

Then there are studies outside establishment-approved radioisotope dating which cast some light on the weaknesses of its underlying assumptions. Young-Earth creationist scientists have dared to use "dating" techniques on materials that old-Earth scientists wouldn't think to try. They used techniques designed for dating rocks assumed to be millions of years old on lava rocks from eruptions less than a thousand years old and found them "dated" as being a million or more years old. They checked for C14 in wood trapped in rocks dated millions of years old, supposedly far too old for any trace to be found, yet they still found some. More recently a project known by the acronym RATE produced some results that are very challenging to the old-Earth view. Perhaps the most significant was the discovery of more helium in certain rock crystals than expected under uniformitarian assumptions. This (together with some other results) suggests that radioactive decay processes which produce helium happened at a much higher rate and much more recently than old-Earth uniformitarianism can allow. Other studies have discovered C14 in diamonds and coal. The old-Earthers have responded with appeals to contamination or effects from nearby radioactive materials, but the YECs have responded to those counter-arguments. At any rate, the old-Earth responses have shown their psychological and philosophical commitment to belief that the Earth is billions of years old.

I must emphasize that none of this is meant to imply that YECs accept and are willing to use the uniformitarian methods as a means of dating the Earth when they happen to fit their belief. Nor do they think that some simple adjustment would make any given technique a reliable way to date the Earth and fossils. On the contrary, all of this is merely for the sake of argument, showing that even with techniques that are aimed at measuring great ages, a lot of examples fall far outside expectations. In the YEC view, the original creation would have had to have some appearance of age (some prefer to call it a mature creation), and then there was a global Flood which may have involved a temporary change in various conditions. Since the Bible is not a science text and therefore not concerned with documenting all the details of the physical processes involved, the accounts of creation and the Flood may leave out a lot of things that happened to the Earth. Both the creation and the Flood would have (or could have) involved unknown "aging" effects, probably varying from one material to another and from one place to another, and at different depths in the Earth.

Evolutionists appeal to gaps in the geologic record to explain away huge holes in the expected fossil pattern, but the holes are systematically where the most important events of evolution would be happening – the greatest changes, presumably taking more time. Usually when fossils are found, they can be identified as known types. There are many fossils of highly specialized types, and vast numbers of fossils of organisms that are neither transitional nor especially specialized. Apart from the expectations of evolutionists, the fossil record is quite representative.

But when it comes to evolutionary expectations, some gaps are rather broad, such as "Romer's Gap," a striking dearth of fossil arthropods and vertebrates right where the most interesting aquatic-to-terrestrial transitional forms should be. Another well-known thin spot occurs just where the transition from apes to humans is expected to have been, although fossils of other kinds of animals are abundant in the same areas and rock layers. Sometimes forms that would serve as intermediate forms are known, but only from rocks that have been dated as younger than those in which the earliest "more advanced" forms have been found. In these cases, evolutionists explain away the fact that the fossils are in the opposite order from their predictions by appealing to "ghost lineages." This is the notion that intermediate forms with a late date show that there were transitional forms that existed much earlier, but those didn't leave any fossils that we've found yet. At any rate, even when there are intermediate forms in the proper sequence, they are usually extremely rare, sometimes fragmentary or poorly-preserved, and almost always too specialized to represent an actual transitional form, but they are said to represent an extinct branch related to the assumed real transitional population.

Evolutionists also need to have several methods for explaining out-of-place fossils. The fossil record is not as neat and tidy as it is presented to schoolchildren and the general public. If some fossils are too much out of place, evolutionists will appeal to "reworking" of old rocks – fossils eroding out of the ground long ago and being buried alongside organisms about to become fossils. If some "later" fossils turn up in much too old sediments, they can appeal to animals falling into ravines, sinkholes, etc., and being buried and fossilized in the older rocks. Animals do fall into caves and die, but with fossils, it doesn't matter if there is an ancient cave structure, when "everybody knows" that fossils of X don't belong in formation Y. Anything that supports this is considered important evidence and anything that can't be made to fit is called anomalous and then ignored.

Fossils once thought to have gone extinct tens or hundreds of millions of years ago have turned up in rocks considered much more recent ("Lazarus taxa"), or as living creatures. Evolutionists lightly explain away all these cases as populations too small to have left any fossils (or so few we haven't discovered them) yet managed to hang on for millions of years, like the ghost lineages. Another way to downplay the significance of such fossils is to claim that the newly-discovered more recent fossil (or living organism) is not closely related to the older fossil, despite the great similarities. Discovering a new species, genus or higher classification rather than a mere variety is also better for a scientist's prestige. Besides, it's easy to find some differences, and claim any similarities are due to convergent evolution. However, take away all the clever, convoluted explanations, and you may see that the numerous cases of fossils appearing in rocks considered much too old for them, widely separated fossils of the same kind, and living animals thought extinct for tens of millions of years or more, all suggest that the vast ages of geologic "time" are illusory.

The cases of "living fossils" and other forms of life which are nearly identical to fossil forms are especially striking. The most famous example is the coelacanth. Coelacanths were a fairly diverse group of fish, the most recent fossils of which are dated at over 65 million years old. Because they have bones supporting their paddle-like fins, they were regarded as possible ancestors of amphibians, given the nickname "old four-legs," and thought to have sometimes crawled onto shore. It wasn't until well into the 20th century that it was discovered that some were still living – not crawling on shores but at great depths. They don't even use their fins for walking on the ocean bottom, though of course their stiff paddling somewhat resembles a walking motion.

The coelacanths are the most spectacular (but not the oldest fossils) of many cases of animals and plants supposed to have been extinct for millions of years that turned out to be still alive. It's hard to believe that relatively small populations managed to survive for such long periods. If we also look at all the organisms that were known all along to be alive and then found to have very similar fossil counterparts dated tens or hundreds of millions of years old, the challenge to the standard geological framework is quite impressive.

Evolutionists also have ways of explaining away vast areas of missing or out-of-order strata. They can say that there was a lack of deposition of any sediments (over vast areas for tens or hundreds of millions of years). They can say that erosion removed many layers of rock (in widespread sheets), and then conditions changed to deposition again. They have to claim that vast tracts of "older" rock were forced to slide up and over younger layers. It doesn't matter if there isn't any good evidence for these processes in a particular case. It is simply inconceivable to an evolutionist that such "anomalies" and "exceptions" could call into question the belief that the general pattern shows vast ages of time passing, no matter how large or how much difference in "age." If anything doesn't fit, it has to be given an explanation that makes it fit, and nobody who wants to be accepted in the establishment dares to look too skeptically at the explanation.

But, despite all these efforts, sometimes accepted dates don't hold up. And not only have single dates been changed, so have ranges of dates and estimates of the rates at which geological events took place. Geological structures which have been given greatly reduced times of formation include: various river-related formations such as the Niagara Gorge (both in the formation of the strata it cuts through and the rate at which it was cut) and the Grand Canyon, caves in general (several individual caves have had their formation period remarkably reduced), the Himalayas, the petrified forest of Yellowstone, the Earth's core, massive granite formations, ironstone deposits in South Africa, and a vast geological formation called the Channeled Scablands, now believed to have formed rapidly during a flood known as the Bretz or Missoula flood, which was larger than any known since the time of Noah.

Most, if not all, of these changes in geological dates and the associated rates of formation have been made by uniformitarian geologists themselves. Not surprisingly, they still don't fall within the young-Earth framework. Uniformitarianism is no longer the ruling paradigm, but all geologists know that if they were to suggest something which strays too far from the general "slow and steady" dogma, their work would not be given any consideration and their career might be over.

This Box Is Strange, Changing, and Full of Holes

Finally, some people (including some Old-Earth Creationists) use "the starlight and time paradox" as an absolute shield against considering that the universe is a relatively recent creation. They may refuse to consider discussing anything with YECs until we explain how light could travel from stars billions of light-years away, in only a few thousand years – and they'll demand that it be done without appealing to supernatural processes! This is rather like arguing that a foil is a better weapon than a saber, and then asking a saber enthusiast to prove otherwise by challenging him to a duel in which sabers are not allowed. However, there are reasons to distrust the idea that the speed of light and other physical constants have always been the same, and there are YEC responses to this problem which indeed don't require special supernatural intervention.

First of all, I'd like to review, in some detail, observations indicating that the natural, modified-uniformitarian model of the universe is not the pure, simple framework it might have been. It starts with the "Big Bang" (the term was first coined as a mockery of the theory), in which nothing existed but pure energy, and the four basic forces of nature had not yet formed from a primordial unified force. What caused the Bang? Where did it come from? Nobody can honestly say we know – some argue that the question is beyond science, and few dare to say there is much we could ever determine by any test. So even the standard model of the universe begins with something that is unlike nature as we know it now, and cannot rule out the possibility that the universe came from a supernatural realm.

For this to be accepted by those who think that matter/energy is all that exists, there has to be some natural model for something to come from nothing without any (known) cause. On a subatomic scale, particles of matter and antimatter appear to pop into existence randomly. Of course, you may have noticed that the universe is a bit larger than a subatomic particle, and these "virtual" matter-antimatter pairs totally annihilate each other again in a very small fraction of a second. So, to make this idea work, the amount of matter-antimatter that just happened to form must have been greater than the difference between subatomic particles and the universe. It had to be so great that some slight imbalance in the laws of physics (known as a broken symmetry) produced a little bit more matter than antimatter, and that small excess is what makes up the entire universe as we know it. The original quantum fluctuation (or whatever it was) must have been that much greater than the universe. How can anyone believe that the universe had such an astounding origin, and still expect everything to have been just the way we see it now for billions of years since then? If we observed fluctuations on large scales in the laws of physics, it would make the Big Bang theory more believable, but then the philosophical assumptions that led to it would be invalid.

But wait! There's more weirdness. It seems that a simple Big Bang would not produce a universe that looks like the one we have. To save the theory, we have to add something called inflation. This is the idea that (for no known reason) the universe must have suddenly expanded very rapidly, faster even than light can travel. This fantastically rapid expansion lasted just long enough to take care of the problem, then (also for no known reason, of course) stopped, and the universe continued expanding at its present rate. More or less. If they can throw in such factors to fix their theory, how can they be so sure there isn't some unknown factor that would totally change our view of the history of the universe?

And we're not yet done with ironing out the wrinkles and patching the holes in what is known as the Standard Model. If we left the theory at this point, there wouldn't be any explanation for why the stars and galaxies aren't spread out more evenly through space. So far, nobody has seen signs of enough matter for gravity to keep it bunched up as much as it is, especially in galaxies. Measurements of the rotation patterns of galaxies indicate they would be flying apart rapidly unless they have more gravity than the visible matter can account for. So scientists believe in Dark Matter, some unknown stuff that has mass and so provides the needed gravitational force, but apparently doesn't have any other known effect on anything. Perhaps there really is such strange stuff out there (gravitational lensing is an observed effect that's good evidence for some Dark Matter, and YEC astronomers believe it exists), but otherwise it sounds a lot like saying that since the theory doesn't work, there must be something out there that makes it work. This is all the more suspicious in consideration that it would take far more of this unseen stuff than there is of stuff we can see!

Then there are a few other strange things which may not be needed to save the Big Bang theory, but they at least show that our ideas about the limits of nature and what the universe is like can change drastically. One of these is black holes, which were first predicted based on calculations of what would happen when a very large star collapsed due to gravity. In some cases, the gravity would become so intense, it would go "off the charts" toward infinity. It would be so powerful that light passing nearby would be curved into it, and no light would ever escape (at least for all practical, non-technical reasons). There are some observations which lend pretty good support for the possibility of these things existing, and if they do, then they are like holes in the idea that the universe is the same all over.

Recent observations have suggested that there is something else very weird compared to our earlier ideas about the way the universe is put together. These observations suggest that the universe is expanding at an increasing rate. This is totally unlike any other known force, and so scientists have added Dark Energy to the mix. This is supposedly a small force that pops up in empty space as space expands, tending to cause more expansion and more space and so more Dark Energy, and so on. There's so much space in the universe now, that all the Dark Energy adds up to more than all the gravity, causing the accelerated expansion, and so there must be more Dark Energy than there is ordinary matter and Dark Matter combined!

And so our picture of the universe has changed drastically since Darwin's time, and even Einstein's time. We didn't even know the universe was larger than hundreds of light-years, with separate galaxies spreading out over billions of light-years, until well into the 20th century. Before the discovery that the universe was gradually expanding, some were suggesting that the universe was eternal. There may still be a few who hold that view. Old-Earth creationists have indicated that any universe with a beginning, even many billions of years ago, is compatible with the Bible.

I recall when it was common to read that scientists were confident the age of the universe was about 20 billion (thousand million) years old, or maybe older. Then observations led some researchers to suggest it might be at the lower end of the range, perhaps just 15 billion years old. As this did not cause too much of a problem or threaten the Standard Model, more researchers began to make such suggestions, so that the 15 billion year figure started to look like the best guess. Once researchers started looking for observations that put the age of the universe at a lower age (and discovering and reporting them helped to get papers published in respected journals), the estimates for the age of the universe continued to creep downward. Eventually this did get to be a serious problem for the Standard Model. Discovering objects that were farther away or that in other ways appeared to make them the oldest thing discovered was also a good way to get published. The estimates for the age of the universe got so low, and for some objects so high, that estimated ages of some stars and galaxies were older than the age of the universe! Some calculations would have put the age of the universe at just a few billion years older than the formation of the Solar System. Eventually everything got balanced out, and the favored age of the universe has ended up (for now) just under 14 billion years old.

So, the things we know we don't know much about in the universe (Dark Matter and Dark Energy) are believed to be far greater (in sheer mass/energy) than the matter and energy we can see. Even the application of purely naturalistic, nearly-uniformitarian assumptions to the study of the universe have suggested a beginning that was far beyond and different from the conditions that we now observe – essentially a supernatural origin. Calculations of the age of the universe have changed drastically. If all that isn't enough to remove confidence that we can safely ignore the possibility of divine creation, there are a number of smaller but still potentially very troublesome problems that might be mentioned. You may not have heard of them all, as indications of problems with accepted theories generally get much less popular media coverage than data that appears to support establishment doctrines like the Big Bang theory.

For example, you may recall hearing about satellite data that was made into a map showing fluctuations in the cosmic background microwave radiation (or CBR) that matched general expectations. That was given some large headlines and a fair bit of broadcast coverage. There was much less popular media attention given to later studies which indicated the data showed a dearth of gravitational lensing, and another which pointed out a lack of shadows from large objects close to our galaxy. As astronomers strive to be the first to discover the most distant galaxies, some distant, "early" stars and galaxies (and clusters and superclusters of galaxies) have turned out to appear surprisingly mature. They include supernovas and indications of black holes, and long strings of galaxies. Astronomers can only shrug and assume that for some reason the universe started taking shape much faster than they expected. Astronomers have also discovered some relatively close galaxies that look surprisingly young, as the universe is supposedly so old that nobody expected galaxies to be forming still. Then there are some mysteries involving red shifts (in layman's terms, a change in the color of stars due to their moving away from us, or more generally the expansion of the universe). Several observations which don't seem to fit the standard theory have been pointed out by people like the non-creationist scientist Halton Arp.

Then there are some things which are just plain odd. They may be given simple explanations when we have more data, but they might suggest further problems with the Standard Model. There is a striking pattern, or at least clumping, of the dates of supernova observations (supernovas are exploding stars), and so far it seems we're a bit short on the count of what's left afterwards (supernova remnants, or SNRs), if stars have been exploding for billions of years. For quite some time, there was a problem with the detection of a kind of subatomic particle known as neutrinos coming from the Sun – there weren't enough of them, by a significant percentage of the predicted count. A solution to the missing neutrino problem was eventually discovered, and some good evidence for it, but it involves the neutrinos having mass and changing, which also hadn't been predicted. This seems to be another surprise that was quietly absorbed by adjusting the Standard Model.

A small but perhaps very intriguing mystery is the Pioneer paradox. The first man-made objects to exit the Solar System, the Pioneer probes were expected to provide new data, but what was completely unexpected was that the measurements of their distances from Earth would become remarkably different from what had been calculated. There may be any number of boring explanations, but this could also be a hint that there is something important missing from our theories about space. Who knows what other little mysteries may be trying to tell us something important? One possibility I wonder about is Hoag's Object, a distant collection of stars that looks like a ring-shaped galaxy with a sphere-shaped galaxy of different stars right in the middle. There are a few other such ring galaxies, and of course explanations have been made for how they got that way, but nobody really knows.

While we're on the subject of cosmic mysteries, the evolutionary story of the formation of the Solar System is rather convoluted. Venus rotates in the opposite direction from the other planets, and Uranus is tilted so far that it looks like it is rolling. The only naturalistic explanations for these exceptions involve powerful forces such as massive objects passing through or near the Solar System, but only affecting one planet each, in just the right way. This seems rather unlikely. It's even harder to accept when considering several other matters that seem to require special adjustments. The orbits or other conditions of several moons are also hard to see as merely coincidental, especially Luna (Earth's moon), Amalthea, and Titan. Luna's size and current distance from Earth provides both beneficial tidal effects and spectacular eclipses of the Sun. The rings of Saturn and other planets have also required special explanations.

Space probe missions to study comets up close have produced surprises concerning what they're made of and how they're put together. More directly challenging to the age of the Solar System is the appearance of comets, which tend to disintegrate or crash into planets after a short time compared to the supposed age of the system. The Kuyper belt, a ring of icy bodies about as far out as Pluto and beyond, might be the source of comets (such as Halley's) with orbits that only go out about that far, but that would also call for a number of cases of massive bodies passing by without doing much else besides tossing some Kuyper belt objects (KBOs) into cometary orbits. There have been some surprises about the size and appearance of the relatively few KBOs that have been seen and studied much. Some comets follow paths suggesting they might never come back. This has led to the proposal that there is another group of icy bodies scattered far out beyond the known Solar System. These so-far-imaginary potential comets are known collectively as the Oort Cloud. Once again, passing celestial objects are required to stir them up and send some our way, to become comets around the Sun. Are these unobserved "passing bodies" a good explanation for all these observed conditions, or just a patch to cover up problems?

Earth itself presents at least one major mystery: it has a lot more water on it (and in it) than current theories can explain. Its formation within the early Solar cloud of dust and gas would have produced a hot, dry ball of crusted-over molten rock. Neither Venus nor Mars has anywhere near as much water. It is well-known that a number of other conditions make Earth "just right" for life. Again, contrast Earth with our neighboring planets. The search for planets around other stars has turned up indications that many stars are orbited by giant planets with orbits that would toss any Earth-sized planets out of the system. As increasingly advanced equipment and detection techniques become better able to detect Earth-like planets, excitement rises as better candidates are found, but often news reports leave out or lightly pass over serious known problems for life-supporting conditions. Keep in mind that any planet the right size and in an orbit roughly equivalent to Earth's would also have to have a proper speed of rotation in the proper orientation. The orbit would also have to be about as round as Earth's, rather than being the same distance on average while zooming in much closer to and farther from the star. The star would have to be much like the Sun (or the distance of the orbit adjusted accordingly) and not too variable, nor a source of intense X-rays or other powerful radiation. The planet would require an atmosphere that was "just right" as well.

Even if a planet appeared to have all the right conditions, maintaining life (let alone fostering the evolution of life) would require those conditions to be maintained as well. Many things could go wrong over the vast ages of apparent cosmological time. Earth is supposed to have narrowly avoided being made uninhabitable by irreversible runaway effects as the Sun changed brightness, the Earth cooled, meteors bombardment it, and something very large smashed into it to create the Moon. Then after life formed, it could have been wiped out, as the atmosphere went through drastic changes in composition, global cooling produced the condition called "Snowball Earth," there were vast and relatively rapid changes in the entire global biosphere, relatively sudden mass extinctions that wiped out large percentages of life all over the planet, and periods of warming that produced moderately warm temperatures even near the poles.

Evolution itself (granting the full history was possible) could have had many different outcomes. Some early form of life could so overwhelmingly win the evolutionary struggle to survive that it would wipe out everything else and either die out, too, or leave the planet covered in this one simple form of life. A number of evolutionists have argued that the appearance of progress in evolution is illusory, because there was simply nowhere to go (besides extinction) but "up" in complexity from the first life forms, and no particular path or even level of evolution was inevitable. There have been some discoveries indicating that the belief that evolution is irreversible is mistaken.

Taking all these things (and more could be added) together, many people find it too hard to have faith that it all came together and life made it through by chance. There may be a literally astronomical number of relatively Earth-like planets in the universe, but the odds against everything being right to support life are still greater.

The young-Earth, global flood view

The presence of the many distinct layers claimed to represent vast periods of time may actually suggest rapid formation – under normal conditions, changes in weather, tides, currents, erosion, stream and river meanders, and bioturbation (the activity of animals stirring up soils and sediments) – all work against the formation of large areas of relatively even layers over extremely long periods of time. Such processes would especially destroy "varves" – fine layers said to represent yearly deposits. Scientists working within the Creation framework have shown by experiments and observations that many layers can be formed rapidly, nearly simultaneously. The supposed correlation of layers in different places may be just another illusion caused by the human propensity to match patterns, or due to common factors other than yearly or other time-related cycles. Geologists generally don't question or try to test their common, fundamental assumptions. Indeed, there's no way to truly test such things, so all they can do is rely on various correlations and hope they are not overlooking something and matching things that appear similar but don't really show the same set of time markers.

The abundance of well-preserved fossils in many locations speaks against formation by slow, ordinary processes. Natural processes such as scavenging, decay, and chemical dissolution normally erase all traces of dead bodies in months or even days. Normally even the hard shells at beaches are ground into tiny bits over time. In contrast, fossils include the delicate bones of birds, feathers, the scales and fins of fish, traces of soft tissues, even the dung of various animals from fish to dinosaurs. (Strangely but perhaps very significantly, hardly any traces of Mesozoic mammal hair have been fossilized.) Very temporary traces such as raindrop marks and footprints suggest extraordinary preservation conditions such as volcanic ash falls or the gradual rise and retreat of waters heavily laden with fine sediment, hardening quickly like cement or filled and preserved with more fine silt.

Those who do not believe in a global Flood claim it would only produce one giant mix of all various life forms, or grind everything to powder, and certainly wouldn't include fossils of delicate things like footprints, but this is an excuse for dismissing the possibilities without giving them any consideration. If the standard view includes explanations for why there are so few fossils before the Cambrian explosion, "ghost lineages," "Lazarus species," places where large sections of the "geologic record" are missing or in the wrong order, surely an alternative view should be allowed to appeal to special local conditions, different pre-Flood geology, surface topology and biogeography, and other such special explanations of its own.

The ratios of various fossils are consistent with a Flood origin – most fossils are of marine invertebrates. The uniformitarian geologists who originally rejected such things as evidence of a Flood and proposed instead that the continents gradually rose from beneath the sea were of course ignorant of many things we have learned since then. The bedrock of the continents is a different kind of rock (granite) than sea floor rock (basalt). While many changes have been made in the understanding of geological compositions and processes, the basic philosophical doctrines remain as passwords to the halls of acceptable geological interpretations.

Erosion happens constantly, almost everywhere, unevenly, and can act much faster than uplift and soil building. There is evidence for vast areas of rapid, catastrophic sheet erosion, but erosion is now concentrated in streams and rivers. Even where the rivers meander through flood plains, they don't leave the smooth sheets of sediments seen in rock strata. The Creation Research Society Quarterly and other journals of creation science constantly present aspects of geology ignored, overlooked, or lightly explained away by the mainstream journals. In some areas, angled strata have been cut off flat at the top, although the strata now are of different degrees of hardness. There are large, high plateaus in western Canada and the USA that are covered with rounded boulders and gravel that apparently came from rocks over 100 miles away. Observations such as these indicate that in the past there were vast, catastrophic erosion events all over the world, unlike anything that has been seen since men began keeping regular written records.

Certain smaller details also suggest rapid formation and thus a young Earth. Polystrate fossils – fossils such as tree trunks that extend through many layers; formations known as clastic dikes and pipes, in which underlying material sticks up into the supposedly much younger material above it, and areas where the layers are smoothly folded into tight vertical curves, all indicate the sediments were laid down rapidly and altered while still soft.

In the case of polystrate fossils, in which fossils (most notably, of trees) extend through many layers, even evolutionists sometimes agree that those layers must have formed rapidly, if only in those areas. They seem to just ignore the observation that the layers look just like those which elsewhere are said to have taken vast ages to form. The dinosaurs were at first thought to have gone extinct due to slow processes alone, such as evolutionary pressures from mammals, but now relatively swift and catastrophic processes are given credit for quickly dropping the curtain on the reign of the reptiles. Similarly, the origins of the modern orders of mammals and birds are now also supposed to have happened relatively rapidly.

Some people are strongly convinced that the Earth is very old by the appearance of large formations derived from organic material such as coal, or minerals like salt (evaporites). The impression they have is based on their habit of, or commitment to, judging things by current conditions and rates of processes. However, it is undeniable that conditions have been very different in the past. Even in the once strictly uniformitarian establishment, it is now understood that rates can vary wildly, and that conditions on Earth have been very different in the past.

Rates of population growth are known to "explode" under special conditions, e.g., algae blooms. The Earth's magnetic poles are known to be wandering and changing in strength, and are believed to have reversed many times in the past. Such changes would have altered other things, such as the levels of radiation reaching Earth, and from that the production of C14. And, if there's anything to the interpretation of various isotopes in ice cores and such, there seems to have been significant changes in global temperature. It's also believed that there have been very great changes in atmospheric ratios of carbon dioxide, oxygen and other gasses. Even those who deny there was a global Flood now accept that there have been great changes in sea level. Huge catastrophes – the Bretz flood, meteor strikes, super-volcanoes – are now accepted by the geology establishment. A divine creation and a divinely caused or directed global catastrophe would totally void uniformitarian interpretations and estimates. Old-Earthers have said since they first appeared that there is no geologic evidence of a global Flood, but that claim is based on preconceived notions of what to look for.

Within the establishmentarian limits, great differences in the biosphere are said to be indicated by fossils in the Arctic and Antarctic regions of dinosaurs and forests. Elsewhere, there are fossils of giant insects and other arthropods, and gigantic dinosaurs far larger than anything walking the earth today. Some of the pterosaurs and extinct birds were larger than what anyone would have believed possible for flying animals. Massive deposits have been chalked up to action of glaciers covering (for all practical purposes) the entire globe, inspiring the nickname, "Snowball Earth."

Arguably, the large deposits of organic materials and minerals are evidence that the Earth has experienced conditions very different from what we've observed, and even beyond the past catastrophes that have been accepted by the establishment. The sheer size and relative purity of various deposits argue for special, rapid formation conditions. Salt deposits, for example, can be so pure that only well-designed facilities for artificial extraction can produce a purer product (*Roads & Bridges*, June 2000, "Knowing What Salt to Buy," pp. 80-81). Many chalk deposits bear little resemblance to the muck on the bottom of seas. The material in coal seams have many differences from the peat swamps they are said to have formed from, and there are no direct observations indicating that such swamps can continue to build up organic matter to sufficient amounts and transform it into hard coal.

As with the candle illustration, nobody really *knows* the true history of various deposits and formations, in spite of all the evidence that seems to show what went on. Often the indications cited as evidence for gradual formation in ordinary ways are weak, or there are clear indications of catastrophic deposition that evolutionists must explain away, such as abrupt layering with no indications of intermediate soils, undisturbed inorganic layers, and those pesky polystrate fossils. Complicating the picture are studies showing that some organic matter can form from abiotic chemicals and processes – methane, simple amino acids, carbon dioxide, oil. It's also been shown in experiments that coal, oil, and other materials can be formed rapidly under the right conditions.

More directly challenging to the old-Earth view are cases of supposedly fragile, relatively complex organic matter preserved in rocks supposedly tens of millions, up to hundreds of millions of years old. Evolutionists hadn't thought of looking for organic material in fossils, as it tends to break down rapidly, and given the vast ages of uniformitarian geology and modern "dating" methods. Some cases were too obvious not to notice. The "oldest" organic materials so far includes pigments in fossils of crinoids ("sea lilies") that are supposed to be over 400 million years old. Perhaps the most astonishing cases were discoveries of living bacteria spores in amber supposedly tens of millions of years old, and in salt deposits dated about 250 million years old. Of course these results have been challenged, but they have also been defended. Ultimately such cases have been ignored, downplayed, and forgotten.

Evolutionists especially didn't expect to find larger and more complex molecules and material such as proteins, cartilage, or any other soft tissues or organic molecules in dinosaur-era fossils. But researchers couldn't help noticing some signs, such as the smell of bone when some fossils were cut, and oily, smelly residues when leaf fossils were released by splitting rocks. Finally, a few began to look closer and do some tests. They were surprised to find organic material such as dried remains of blood, and bits of tissues that were still flexible, in fossils as old as dinosaur bones. Of course they couldn't allow themselves to believe that such things indicate there was something wrong with the assigned dates, so they have put forth theories they hope will explain how such things could endure for such incredible lengths of time. In areas such as physics or chemistry, such anomalous results would incite calls for re-evaluation of fundamental assumptions, but the beliefs being called into question by these discoveries are too important to the worldview (and careers) of those who hold them.

Another entire class of data that weakens the case for Earth being billions of years old is ratios of various elements in the atmosphere and the seas. Of course, the rates at which such materials are added and subtracted are difficult to measure and the net gain or loss calculations can be rather complex. Balances can rise and fall in cycles. Still, there is a surprising number of cases in which it appears concentrations are still building up, with no maximum in sight. Perhaps the best example is the sodium content of seawater. Sodium is an element found in common salt, and it is relatively easy to measure salt content in rivers and estimate how much salt is entering the oceans at the current rate. Old-Earthers have suggested some ways in which the salt might be getting removed, but those are highly debatable. When we consider that the ocean probably had a fair amount of salt to begin with, and that various catastrophes and global environmental conditions would have greatly increased erosion rates and the inflow of salt, it's a wonder that after supposedly billions of years, the ocean isn't full of all sorts of minerals. Billions of years? We might wonder why the whole Earth isn't covered with a soupy mud.

However, this is also like dating a candle by its burn rate, when you don't know what its original state was nor what might have happened to it before the present. Creationists only bring up such considerations to point out things that don't fit billions of years of things being the same. For those who have faith in the consistency and regularity of nature, it's easy to dismiss such warnings. In the first place, they undermine the comforting faith that studies of the past can be just as reliable as any area of science. It's also easy to feel that supporting evidence "makes the case" and any exceptions merely "prove the rule" – especially when expressing doubts can ruin your career. It's easy to laugh at suggestions such as "God could create old rocks." Most young-Earth creationists reject that possibility, comparing it to the notion that God created fossils to test our faith.

While we're out on the fog, let's think out of the box

But where do we draw the line once we've stepped outside the proper limits and method of science, and started to play the evolutionists' game, but allowing (and insisting on) Divine Intervention? How can we tell what God might or wouldn't possibly do? What sort of things might a creator put in a world made to sustain a rich panoply of life, and to nourish, entertain, and educate brilliant people from the beginning? And consider that the people would be told by the Creator that the Earth and the whole universe had been created miraculously just a few days before, so they would have no reason to think of trying to discern its age from examining it. If you make the effort to seriously consider the matter and dare to step outside the box a bit, the picture may look different from what anybody has been teaching.

Take, for example, an extreme possibility, that God would create "old" rocks. Can't we think of any reasons to create "mature" ores? Can we be sure we have considered everything that God would have taken into account? Certainly it would be no problem for God to wait billions of years for Earth to age naturally, but if He chose to create it "ready to move in" for humans, it certainly wouldn't be as if it had just naturally formed into a solid ball from space dust and gas! Any world created to immediately support human life would have to have a number of features that would make it appear "old" if judged by naturalistic assumptions. Throw in considerations of safety, education, and who knows what other needs and purposes that might be involved in the creation of a world best suited for complex, intelligent beings, and it is obvious that we can't be sure of what various things may have been different from our natural expectations. Keep in mind that God did not create the world in a way meant for us to figure out its history, and God was there with Adam and Eve, to reveal to them His series of more-or-less instantaneous acts of supernatural creation. Then, too, there's no telling what special alterations to the Earth may have occurred after the creation, especially at the introduction of the Curse of Sin, and during the global Flood.

It might seem odd or just plain wrong to think that the forces of nature may not have always operated as they do now, or that they varied temporarily, or from one place to another. Supposedly, science requires a complete uniformity of natural laws. However, even within the naturalistic philosophy approach, it is now believed that the universe started with very different conditions, and there are places in the universe where conditions become so different that we cannot detect anything beyond a certain limit. If the entire universe sprang into existence as a random fluctuation in nothingness or a background quantum field, why shouldn't we expect fluctuations within the universe, alterations in space and time that would be unpredictable in their specific effects? If any variations can be established, especially if a small detected change could be shown to imply a much greater change in the past, the foundation of the establishment's view might be weakened too much to ignore. Not surprisingly, the few investigators who dare to look into such matters from inside the establishment tend to tread very carefully, and even slight departures are vigorously challenged.

We must ask: Granting there may be reason to doubt the conventional story of the formation of the universe, what exactly do Young-Earth Creationists propose instead? What about that starlight and time problem? There are actually a number of possibilities.

I personally believe, in the first place, we should consider that this is one case in which all we can certainly say is "God did it." Creationists actually like to find natural explanations for everything possible, while knowing by faith that God is working behind the scenes in all things. At some point, however, either some obvious divine (and otherwise inexplicable) intervention was required, or the universe could have formed without God. How could a universe that looks like it formed naturally, without God, show God's handiwork, power and glory?

Everybody knows that science can't prove or disprove the existence of God (or anything else supernatural). Philosophy and logic tell us that we can't rule out the possibility that the universe sprang into existence just a moment ago, complete with our memories of having lived up to now. There's just no reason to believe that's the case.

However, we do have a history of belief in divine Creation with humans included at the beginning, and many Christian theologians, philosophers, and scientists giving reasons for its having occurred about 6,000 years ago. A believer in the Old-Earth view might mockingly say that it requires God to create fossils in the ground, artifacts, monuments and ruins that no human built, and many people with virtual memories, but that idea is derived by cramming the made-up "history" of civilization into the Young-Earth time limit.

In the YEC framework, only two humans were created directly, and while it's reasonable to expect they were granted great intelligence, there's no indication that Adam and Eve were created with memories. If written historical records extended back many thousands of years, belief in a history of less than 10,000 would truly be unreasonable. Given that current theory and dating place anatomically modern human origins over 300,000 years ago, and symbolic markings at 70,000 years ago, we might well expect to find historical writings dating back 20,000 years or more – but they don't, so this is not a problem.

All YECs, by definition, hold that at least the Earth was divinely created in a complete form, that is, with some appearance of age, or as many prefer to say, it was a mature creation. Some hold that the raw matter of Earth was created billions of years ago, but the world as we now know it was formed (or re-created) a few thousand years ago.

However, while most accept that the Earth was created with its present atmosphere (or something a good bit better), a complete set of living things, rivers flowing comfortably between banks, sandy beaches, and many other features which couldn't appear (by natural processes) in six days, most have a problem with starlight being created already coming to Earth. Taking a two-book view, they feel it would be like God was writing lies, as the light would have to portray a virtual history of stars and galaxies that never really happened. For some of us, it is enough that God has revealed (and could have explained to Adam and Eve) that He created light separately from the stars – days before them, in fact. When He said "Let there be light," He could have called into existence all the light a full-blown universe would have, and sustained its ongoing creation until He formed the stars to provide the natural supply.

However, for those who are uncomfortable with the idea that the appearance of long-past conditions and activities of distant stars is just a divine art form, there are an indefinite number of other possibilities, with two or three being currently more or less popular and subject to ongoing investigation and development.

For one thing, God may have created the stars and galaxies whole, but without the physical constants like the speed of light at their present values. Light might have filled the universe in an instant, and later slowed to its present speed. The leading proponent of this approach (perhaps not exactly as I've presented it) is Barry Setterfield. The speed of light is related to a number of other physical constants, and theoretically a different speed of light would also have effects on other things, but if we're wrong about the speed of light being constant, we can't say what else might have been different in the distant past.

A more recently proposed possibility is sometimes known as the White Hole cosmology. It was proposed by physicist Dr. Russ Humphreys. This approach takes advantage of modern knowledge that gravity and time are tied together. A White Hole is, in theory, something like the flip-side of a Black Hole, with extreme gravity but spewing everything out instead of dragging everything in. If God created the Earth at the center of a White Hole, the gravitational difference or distortion between the Earth and the things getting spread out to the far corners of the universe would also mean that time was passing at different rates. In this model, the outer reaches of the universe are in fact billions of years old, while Earth is only a few thousand years old.

If these ideas don't pan out, or just don't suit your fancy, they hardly exhaust the possibilities. Waiting in the wings (from both materialists and creationists) already are several possibilities people are just beginning to look into. Some have proposed that the speed of light changes with direction in a relativistic way that wouldn't show up in previous tests. Perhaps its speed only changes over cosmic distances. Nobody knows how the puzzle goes together, and we know there are lots of missing pieces, and some pieces we have don't seem to fit.

And so we conclude our review of the philosophical shift in the modern concept of science, how it prepared the way for the theory of evolution and evolutionism, and how it continues to rule certain areas of science. As previously mentioned, it would actually be many decades before the last pieces would fall into place and evolution became generally accepted, but as long as the new, naturalistic philosophy of science remained in place, there was a gaping hole that could only be filled by some naturalistic theory of evolution.

Aside from expanding the number of jobs for professional scientists and enlarging science textbooks, however, this modern philosophy of science has had little effect on science. Naturalistic explanations of the origin of the universe cannot produce any practical benefits. Evolution is supposed to be the foundation and cornerstone of biology, but biology was studied long before evolution was widely accepted, and Darwinism hasn't produced any useful results that are inconsistent with a creation framework.

In the final part of this book, we will look at the final steps leading up to Darwin's falling into the last gap in the scheme of philosophical naturalism, the rapid acceptance of evolutionism, and the many stumbles the theory has had since then. Then we will examine key details of the theory, from those things which are accepted on faith, to those which are mostly matters of interpretation, to things which creationists agree are entirely factual but are commonly given a veneer of evolutionism. At that last point, we will have returned to the sort of scientific investigation that science began with, and which has always provided all the practical benefits of science, and provides the only good possibility of common ground and helpful dialogue between creationists and evolutionists.

Part IV: Dogmas – Incompatible Worldviews Fuel Endless Debate

Chapter 10: The Heart of the Debate

We all want certainty and security in a mysterious world – one way or another

Some people think that the debate is a matter of ignorant religious dogmatists stubbornly resisting scientific facts. The fact is that there's dogmatism on both sides, the significance of facts are interpreted by the dogmas, and some things considered facts are really only beliefs and opinions. It is a strong human tendency to want the world to be as easily understood as possible, to explain everything within one tidy mental gift-wrapped box. Our deeply held philosophical and religious beliefs provide that security blanket, but at the cost of making it very difficult to be objective about things that don't fit.

For centuries, this comforting social and mental consistency was provided to Europeans in the form of Christian dogmas. Increased travel and exploration exposed Europeans to other worldviews, and as science developed, it seemed to create a clear distinction between the natural and the supernatural realms. At first, most Europeans, including scientists, simply adjusted their views a bit to fit where they had to. Others, however, began to conceive of a whole new framework for viewing the world. The Christian worldview saw the natural world as a small, temporary part of reality, within and sustained by God's greater, supernatural realm. The new view started by picturing the natural realm as totally separate from the supernatural, and ended up saying that the natural realm is all that exists. At a crucial period in history, it began to seem to many people that everything could be very simple indeed. There was hope that science would soon discover the answers to all the big questions, solve all the last major mysteries, and wrap everything up in a tidy package that nobody could doubt or deny.

The world is not simple, though. We now know that things aren't that simple. Mysteries remain and new mysteries and paradoxes have come into view. Most of this is normally beyond our ability to perceive, although it surrounds us, from the subatomic scale to the vastness of space, we have discovered many strange things we still don't understand, and indications that we know very little about what most of the universe is made of. What we do know about the world, even at its tiniest scale, has turned out to be far more complicated than had been imagined.

When people sense that something comforting is slipping away from them, they often fight hard to hang onto it. Evolutionism is a comfort to many, who like to think it is merely accepting scientific facts. They deny there is such a thing as Evolutionism. This is the heart of the debate, its origin, and the reason it has not gone away, but rather is more lively than ever. If Evolution (in its full, common sense) were truly scientific fact in the same way as gravity, it wouldn't still be doubted or denied by so many intelligent laymen and scientists. It is the scientific and logical reasons to doubt it which provide the Creation-Science and Intelligent Design movements with so many scientists and well-educated amateurs.

It's a fact that the details of Darwin's proposals for the specific mechanisms of how evolution worked were discredited soon after he proposed them. Likewise, nobody can demonstrate that genetic variations and natural "selection" have the power to produce all forms of life from some minimal common ancestor. Since Darwin's day we have learned much that supports the position that living things are dynamically complex in a way that only intelligence can account for. We've also learned that variations are fixed (Mendelian genetics), or pre-existing (epigenetically-controlled) options, or degenerative; never both entirely new and beneficial through increased complexity. While evolutionists still claim they are closing the gaps between different forms, creationists are learning the biological significance of the differences. Creationists and evolution-doubters are growing in number and starting to have the resources to do significant original research. However, many Evolutionists continue to oppose viciously anyone who departs from their beliefs. Clearly, believing in evolutionism provides a number of personal, philosophical, psychological, and social benefits, and so anything that threatens it arouses extraordinary resistance.

For one thing, there's the comfort factor. It's obviously comforting to believe that we do not live in a "demon-haunted world," as Carl Sagan illustrated (by his book, *The Demon-Haunted World: Science as a Candle in the Dark*, copyright 1995). Although religion (monotheistic religion, anyway) is said to provide the ultimate simple answer, "God did it," God is an infinitely intelligent being with complex and incomprehensible motives and purposes. According to the Bible, God reveals that "My thoughts are not your thoughts" (Isaiah 55:8). While many find comfort in the Bible and what God has revealed to our limited understanding, many others do not like the idea that some mysterious and paradoxical things are forever beyond the understanding of mortal Man.

In contrast, the naturalistic philosophy view allows the maximum possible degree of simplicity and comprehensibility. There is no "beyond," there are no complex and mysterious supernatural beings with plans and motives beyond our ability to know. The matter/energy that we are made of is *all* that we are, and nothing very different exists anywhere. Geology can be explained by the very same forces we see today, operating at roughly the same rates. The chemistry of living things is not significantly different from the chemical processes outside of them. The variety of living things is no more wonderful than the variety of a litter of kittens, simply repeated countless times. There is hope that we can someday understand all that exists in terms of interactions equivalent to billiard balls bouncing off each other.

However, all that is a matter of past theory. Those *were* the beliefs and hopes, and some still cling to a worldview only slightly different. Those who know the latest research in each area will recognize how things have changed, even within the naturalistic philosophy. All too often, however, experts in one field who know the old ideas there have been discarded are unaware of the scrapping of old ideas in the others, and so are reassured that the changes in their own field are not too significant.

There must be a lot of comfort and pride in being one of the Enlightened Ones. And certainly it's satisfying to believe our efforts have revealed the true story of the ancient past (and likewise can disregard the Scriptural revelation that we will stand before our Creator and be judged by Him – II Peter 3:5-7). The prestige and pride of being accepted among professionals are powerful motivators. They probably played an important role in introducing naturalistic philosophy into the field of geology.

As other areas of science advanced (in harmony with religion), men who studied geology may well have felt hampered by the limitation that the Bible placed on their field through the account of the global Flood, desiring to go beyond all the work to be done in describing the layers and formations in the earth, the types of rocks and where they occur, etc. Exactly how they fit into history with the Flood is a large area for research in itself. Early on, however, some researchers wanted to do more than catalog, describe, and interpret the evidence within a Biblical framework. Some rationalization was needed for coming up with explanations entirely on their own. Deistic theology and naturalistic philosophy provided this excuse. Natural human observations and logic led to expectations that everything will continue as it has, and has always been the same. This is the basis of uniformitarianism. If everything always happened as it does now, discerning the past by studying traces observed in the present would be relatively simple.

On the other hand, since the Flood was a divine intervention, it's quite possible that a proper understanding of geology is not achievable by such means alone. Naturalistic philosophy is the only framework which makes it possible to expand purely "scientific" work to include comprehensive explanations about the unobserved formation of geologic features.

Even scientists who are creationists sometimes seem uncomfortable with the idea that they may have to rely on a supernatural explanation, that something in "the book of" nature may simply not make any sense apart from the light of revelation of something more. This can be a problem in some areas of astronomy and geology. The spectrum of positions between YECs and fully atheistic evolutionists is determined by what limitations on human observations and logic different people are willing to accept, their desire for scientific prestige or social comfort, and how willing they are to accept the Bible as a divine revelation with clear historical information on the origin and history of the heavens and the Earth. (It bears repeating that nobody, not even the strongest YECs, regard the Bible as a science textbook, let alone one that makes geology, or any other area of science, a waste of time.)

Perhaps the most practical, personal, and powerful reason for staying within the good graces of the naturalistic philosophy-based scientific Establishment is protection of one's professional career. While there isn't a secret cabal of scientists plotting a conspiracy to enforce evolutionism, there are powerful institutions and individuals controlling access to expensive equipment, publication in journals, and funding. Perfectly loyal evolutionists who have simply suggested unusual variations, or criticized the methodology of an evolutionary study, have been chastened back into harmony by editorials denouncing them for giving comfort to creationists, or worse yet of being closet creationists themselves.

In debates, evolutionists have claimed that anyone finding proof of creation would become famous and given a Nobel Prize. This is unrealistic. In the first place, such things are not subject to outright proof, and scientists generally agree that their work is not about producing such proofs. In practice, what researcher would dare engage in research that might cast doubt on evolution, and what journal would publish it? With the naturalistic philosophy framework now endorsed (in practice) as the essential part of scientific endeavor by all the major scientific journals and institutions, anything that challenges it is automatically rejected as unscientific. Dr. Jerry Bergman has collected over 1,000 cases which make it clear that evolutionists are willing to destroy careers first and not bother to ask questions later, rather than risk a weakening of their monopolies in academia, government-funded institutions, etc. (Read *The Slaughter of the Dissidents* by Jerry Bergman and Kevin Wirth, Leafcutter Press, 2011, and their more recent (2016) *Silencing the Darwin Skeptics.*)

Evolutionism has done nothing to advance science, it has only added a lot of stories which can never be verified. The few practical (or at least verifiable) advances attributed to it are also compatible with creationism. However, the philosophical and personal advantages that made the underlying worldview attractive in the first place have sustained the 19th-century notion into the 21st.

I will next show how the adoption of this worldview was completed when biology was infected by the naturalistic philosophy of geology, how it's triumph has been challenged by the rise of Creation-Science and the Intelligent Design movement, and in the last chapter, finally discuss where all this leaves us with our beliefs and the facts that we now know.

Chapter 11: The Origin, Rise, and Stumbling of Evolutionism

Section I: The Cradle of Evolutionism

Having set the final pieces of the historical stage upon which Charles Darwin was thrust, I will depart from a strictly chronological progress. I believe Darwin's work is best understood by viewing his century as one great historical drama made up of many parallel waves of new ideas and attitudes. Darwin was swept along by these waves like a piece of driftwood until he came into the spotlight. His ideas were quickly accepted (though carved up to suit) so that he could serve as a figurehead for the ship of Evolutionism which continues to ride the wave of scientific triumphalism. Not surprisingly, the history of this historical science involves matters which are still relevant and arguable. First, a brief review is in order, as one of the men most responsible for helping Darwin catch the wave and grab the spotlight, Charles Lyell, also established the stage and backdrop for Darwin's grand drama.

Subsection A: Lyell and the Triumph of Uniformitarianism

We've seen that the history of evolutionary ideas and theories began long before Darwin. In fact, everything was in place before the end of the 1700s. The essential philosophical foundation, that is, the uniformitarian materialistic framework, was widely accepted. Ideas very much like Darwin's had already been set forth. Evolutionism might easily have staged its coup seventy or a hundred years earlier. The government that came out of the French Revolution, that fruit and crown of the "Enlightenment," might have established evolutionism as the reigning doctrine of biology, but it soon collapsed in a blood-soaked suicide. Furthermore, the masses of common people, and most prominent scientists as well, still rejected evolutionism as going too far from the plain message of the Bible, and also lacking observational support. In other areas of science, researchers required the sort of repeatable demonstration, or at least observation, which evolutionists were unable to provide.

In geology, however, a new idea of what science was all about was coming into power. Geology had gotten off to a good start in the 1600s, with men like the biblical creationist Steno. Like most sciences, geology was more of a hobby than an occupation well into the 1800s. Also, major discoveries of basic facts were still being made near the end of the century, so that Darwin and others would note that the field was still in its infancy.

However, in another sense, "modern" geology was devised during the late 1700s and established early in the 1800s. Geology hadn't progressed in the way some other sciences had. The late-18th-century concept of uniformitarianism that Hutton promoted provided hope for making geology more than descriptive. It was accepted on the basis of belief that physical laws (and thus natural processes) were divinely ordained and unalterable. The practically atheistic consequences of this belief were easy to overlook and took effect gradually. With James Hutton leading the way and John Playfair providing a popular-level version, the study of the Earth became dominated by faith that present natural processes had never been interrupted nor greatly altered.

This injection of faith was made acceptable by two popular notions: that nature was like a book that trained men could clearly read, and that the Creator had so established the laws of nature that He Himself would not make any exceptions. There is a subtle distinction between the "natural philosophy" that focused on observable, demonstrable operations of nature, and the naturalistic philosophy that assumes nothing exists except such natural operations. Thus, hardly anyone seems to have noticed that the investigative tool taking shape as science, which previously had concerned itself only with what was directly observable and preferably experimentally demonstrable, had been infected by a parasitic offspring of religion and philosophy.

In the early 1800s, the uniformitarian view of geology became thoroughly established and entrenched. Charles Lyell was the new champion of this view. Lyell's training in law helped him write persuasively in favor of uniformitarianism, and his appointment as head of prestigious science organizations allowed him to complete the blockade of all opposition. A casual historical sketch might easily portray Lyell as the Father of Modern Geology, having taken the rough ideas of Hutton, applying mounds of geological facts, and winning over a large number of doubting geologists. Historians of science might say he was merely the most famous of a number of uniformitarian geologists of the time, but Lyell did play a major role in locking uniformitarianism in place as the established and unquestionable fundamental doctrine of geology. He also was largely responsible for inspiring Charles Darwin to develop his ideas about evolution, and to publish his ideas, although Lyell himself was reluctant to accept the full extent and implications of Darwinism.

It must be stressed that Lyell's uniformitarianism was not new, was not established by facts, was not atheistic, and was based on philosophic and religious views. It is true that Lyell cited many geological facts in his *Principles of Geology* (in three volumes, the first published in 1830), but they were presented as evidence for uniformitarianism by the application of estimates which assumed an extreme and unrealistic degree of uniformity in natural processes over time. They were generally accepted at the time because of the previously established popularity of the view of nature as a regularly-running machine. This view was largely promoted by people who believed in God, but not in the reliability of the Bible to accurately reveal truths about God's supernatural creation, and judgment of the whole Earth in Noah's day. While geologists who believed in a relatively recent creation and global flood continued to publish works arguing that various facts supported Flood geology, the men who controlled geological associations and their journals either ignored these works entirely, or attacked the writers without considering their arguments.

It was only natural that a believer in uniformitarianism who had argued so persuasively for it should rise to the top, although some thought he took the idea too far. After Lyell was made president of the British Geological Society (1835-37) and (in 1838) president of the geological section of the British Association for the Advancement of Science, his extreme form of uniformitarianism was established as virtually unquestionable for a hundred years or more. Whenever newly discovered facts challenging the view were presented in the late 1800s and early 1900s, almost all geologists responded by closing ranks and defending Lyell's philosophy as if it were a sacred doctrine. After all, they had been thoroughly taught that their science depended on it.

Uniformitarianism was established so thoroughly in Darwin's day that the door was wide open for a theory that similarly framed biology in terms of the slight changes from one generation to the next that we can observe. Strict uniformitarians with deistic ideas, including Lyell himself, preferred to believe that there hadn't been any significant changes in living things since a creation lost in an immeasurably distant past, but Darwin's theory was the biological incarnation of the same philosophical method of explaining the past with nothing but ordinary natural processes, and it eventually won over almost all geologists and biologists.

Subsection B: The Time Was Ripe for Darwinism

There are two historical facts which are very helpful in fully appreciating why evolutionism became widely accepted when it did. The first fact is the spread of naturalistic philosophy and belief in uniformitarianism, with its vast ages in the past. The other fact is the generally modernist, progressivist, scientific spirit of the age.

For many countries, the 1800s were a cauldron of strikingly new ideas in every area of life, and a flurry of scientifically-derived technical achievements that seemed well on the way to producing a real-life Utopia. As Francis Bacon had predicted, the success of "experiments of light" (basic research) led to "experiments of fruit" with practical applications – technology, engineering, medicine. Science had begun at the same time that Europeans began sailing to the far corners of the world, and now it was changing the world at the same time that European and American explorers were trekking across the last major unexplored (by "white men") regions of the continents. Although the successes of science were also accompanied by excessive enthusiasm and frauds, the glory of science had never been so great as it was then, and it wasn't until almost the middle of the 20th century that its reputation became seriously tarnished.

Consider that the Lewis & Clark expedition took place in 1805, and by the end of the century the "wild West" was already legendary. The Pony Express had been formed and soon replaced by the telegraph and by trains delivering mail across the continent. Likewise, the interior of Africa was largely unknown to Europeans at first, but men like David Livingstone brought to light much of "darkest" Africa. Ocean voyages were at first made entirely on wooden sailing ships such as the HMS *Beagle* which Darwin boarded in 1831, but eventually iron ships with steam engines began replacing the older vessels. The war between the Union and the Confederacy in the 1860s included the first ironclad warships with rotating gun turrets. In many positive and some negative ways, science affected every level and aspect of society, including the arts. Daily weather forecasts became popular additions to newspapers. Even the term "scientist" was coined in this century. The full story of the increase of scientific and technological developments in 19th century is a subject worthy of many books. For this overview, perhaps a partial list will suffice.

The greatest and most obvious fruits of science were seen in mechanical and electrical technology. Inventions made major changes to life at home. This was the century that would eventually see the invention of the indoor flush toilet, typewriters, sewing machines, vacuum cleaners, practical electrical lighting, electric fans and clocks, and the first automobiles.

Communication over long distances was revolutionized, first by Morse Code and the telegraph system by which it was first transmitted, including the first telegraph cable laid across the bottom of the Atlantic ocean to carry messages rapidly from one hemisphere to the other. Then, the invention of the telephone allowed people to talk directly to each other over long distances. Photography was developed to the point where amateurs could cheaply and easily take part with their Kodak box cameras. The phonograph and early forms of movies provided entirely new means of entertainment.

While very few people envisioned where such things would lead, early adding machines, Charles Babbage's mechanical computer designs and experiments, Boole's work on binary logic, and Hollerith's punch-card data system were forerunners of our modern computers. Likewise, early submarines, automobiles, and failed attempts to build powered aircraft may not have seemed impressive at the time, but they provided hints at the end of the century that greater things were still to come.

The developments in fundamental physics and chemistry research may have had little impact on the citizens of the 19th century, but they, too, presaged amazing discoveries and inventions still to come. The first modern forms of atomic and molecular theories were proposed and eventually accepted and developed. Organic chemistry was developed and industrialized. Mendeleev contributed his periodic table of elements. Advanced measurements of the speed of light were made. Investigations into Brownian motion and thermodynamics, along with the discoveries of X-rays (1895), electrons (1897), radium and polonium (1898) set physicists on the road to the 20th century's mind-boggling changes in our understanding of fundamental aspects of matter, energy, space and time.

Of course, advances were made in biology and medicine as well. The universal appearance of cell structure in living things was proposed. The presence and importance of sub-cellular structures began to be investigated. The specialized cells required for reproduction were discovered and their operation studied. The Germ Theory of disease was proposed and bacteria were studied, leading to the introduction of antiseptic surgery. Some viruses were discovered before the century ended. Early scientific studies of nutrition would lead directly to the 20th-century understanding of the importance of various vitamins and minerals.

The boom in science and technology, however, is only part of the picture of the spirit of the age which led to the acceptance of evolutionism. The early form of materialistic uniformitarianism was just one of many philosophies which became popular in Western society, along with a number of more-or-less new religions or religious philosophies. The new movements include Mormonism; Millerites and Adventism; Church of Christ, Scientist; "Agnosticism" (a term coined by Thomas Huxley in 1870); Spiritualism; Baha'i; and Theosophy.

Meanwhile, anti-biblical views dating from the "Enlightenment" or earlier were promoted by a number of religious modernists. Wellhausen and others promoted ideas harking back to speculations by Jean Astruc and David Hume in the 1700s, proposing that the first books of the Bible were not written by Moses, but by a hodgepodge of later writers and edited together by still later scribes only a few hundred years before Christ.

Along with the changes brought about by new technologies and the introduction of new religious ideas, a number of new social movements began in the 1800s. The first were largely religious, and later at least optimistic, but as the 20th century approached, many took a darker tone. Early on, some of the first modern missionary and evangelical societies were formed. Abolitionism led to the end of slavery in Britain and eventually in the United States of America. The ill-fated temperance movement began and grew. Both the YMCA and the Salvation Army were formed. The first steps towards giving women the right to vote were made.

The Romantic movement's reaction against Rationalism contributed to experiments in social Utopianism. Meanwhile, Rationalism contributed to Utilitarianism and Pragmatism. All three of these were well-meaning attempts to provide new, modern frameworks for society, replacing the traditional, Biblical basis for morality and social concerns. The Utopian communities generally fell apart after a short time, having run into the unpleasant realities of human nature. The other two movements were equally unrealistic, and their cold, materialistic view of human life produced more harm than good, as illustrated in the works of Charles Dickens.

Perhaps the dominant social force of the period was *laissez-faire* capitalism, the belief that business ventures should be totally free of government regulation. The excesses of uncontrolled capitalism caused a reaction that included the pseudo-religious Social Gospel movement, an attempt to keep a strong sense of purpose in churches that had largely given up belief in traditional Christianity. The belief that it was best to allow businesses to economically fight it out, so that the best would rise to the top, may have influenced Darwin, and likewise Darwinism would be used to justify this economic struggle for survival. Similarly the success of British Imperialism in reaching its peak, and the widespread colonization of Africa by Europeans, may have provided Darwin with a mental framework for seeing a process of survival of the fittest in nature. Later, proponents of European dominance of the world saw justification for it in Darwin's work.

However, while Darwinism was used to support these "Right Wing" movements, it was Marxism that first officially embraced it as the key to history and the guide to the future. Karl Marx saw in it the natural, scientific justification for changes in which new systems replaced the old. Communist countries have also embraced evolutionism. The Soviet Union espoused its own sort of fundamentalist alternative Darwinism known as Lysenkoism, after its leading proponent, Trofim Lysenko. More directly, Darwinism led to Social Darwinism (which may include the other movements that saw Darwinism as supportive), Eugenics (begun by Darwin's cousin, Francis Galton), and various forms of Sociology. It would eventually also be favored by Hitler and the National Socialist party (Nazis).

There were also one or two new movements in art and literature which were not as dramatic or important in their effects on society, but help to illustrate how deeply the Western intellect had been affected, and changed from old ways of looking at the world. The 1800s saw the beginning of Modern Art, especially in the form of impressionism. For the first time, artists willfully neglected efforts to reproduce the physical appearance of their subjects, in an attempt to express their feelings about them. This would eventually lead to the various forms of abstract art of the 20th century, in which there was no attempt at all to depict a physical object.

The 19th century also saw the beginning of a new form of literature. In 1818 Mary Wollstonecraft Shelley wrote *Frankenstein*, inspired by tales of scientific experiments involving electricity and the muscles of dead animals. Later, Edgar Allen Poe would include scientific aspects in one or two of his horror stories. By the end of the century,

Jules Verne and H. G. Wells became the first major authors to write a number of books in the genre of science fiction.

It should be noted also that some writers joined with the public in feeling some reaction against the scientific craze. This can be seen in Walt Whitman's poem, "When I Heard the Learned Astronomer." Certain passages in the works of Charles Dickens and a short story or two by Samuel Clemens (Mark Twain) make fun of the tendency to explain everything in terms of scientific observations and discoveries of new phenomena. Some entertainers and writers weren't content to borrow from, or lightly make fun of, the scientific enthusiasm. Hoaxes based on the public's acceptance (gullibility) towards science had been around since the time when scientific wonders seemed like a new form of magic. The showman P.T. Barnum displayed some of the best examples, such as the "Fiji Mermaid," a sewn-together work of taxidermy combining a monkey and a fish. Edgar Allen Poe, Mark Twain, and other writers and journalists produced reports of astounding scientific discoveries that seemed (and were) too fantastic to be true, and yet apparently fooled a number of people who should have known better.

To complete the picture, it should be noted that there were one or two areas of research that were bad examples of "science," such as Phrenology, the attempt to discern people's mental tendencies and capacities by studying the outer shape of their heads. Hypnotism had always seemed on the border between science and magic, and it was applied in the 19th century to treat mental disorders, with more speculation and philosophy than hard science. This eventually led to "psychoanalysis" and the works of Sigmund Freud and Carl Jung, much of which have been discarded by modern psychologists.

Subsection C: The universe as a box of billiard balls

We can see in this overview of the 1800s a major factor in the acceptance of Darwinism: more than ever before, it appeared that we were on the very brink of discovering all there was to discover about the world, through the power of science. On top of that, more than any time before or since, it appeared that the universe could be explained in terms of natural forces just like the ones we experience every day. This appearance gave birth to a myth which is still one of the major factors holding people back from accepting creationism. The myth is that science has been steadily pushing back the boundaries of the unknown, and thus leaving ever smaller gaps in our knowledge of the universe. According to this view, as these gaps steadily close, everything will be understood in terms of purely natural forces, so there won't be any reason to believe in God (or anything else supernatural). Thus, any belief that something more than natural forces was required to produce life in the universe is chided as a "God of the gaps" faith. It is said that anyone holding to it is clinging to the last tiny bit of the unexplained, which will also be explained soon.

Young-Earth creationists are probably standing stronger against this belief than anyone, and yet I think we're still sensitive to it. Several areas of creationist research seem to strive toward providing more natural explanations for aspects of the creation of the universe and of the events involved in the global Flood. The more I've thought of it, though, the more I think that there's plenty of solid ground to stand on within the "gaps" – by some measures, there's far more that we don't know than we do. At any rate, it seems odd for believers in God to reject faith in the direct work of God in creation, when the alternative to faith in the so-called "God of the gaps" is to just roll over and settle for a god who didn't do anything except what could have happened without him.

It's also a shame to give up faith when this idea of the shrinking of the gaps, the triumphal progress of science in all areas, is a myth. It may have seemed true in Darwin's time, but we should know better now.

As the 1800s progressed, things looked like this: The idea that "the Church" in general had strongly opposed the geocentric theory and the myth that it had opposed belief in a spherical Earth had been promulgated before the middle of the century. In geology, the extreme uniformitarianism of Lyell seemed undeniable. Lyell, like Hutton before him, claimed that for all he could tell, the Earth had no beginning and might never end. Not only did he reject a relatively recent divine creation and a global flood, he denied that there had ever been any catastrophe larger than had been recorded in recent history.

To Darwin (and many others at the time), heredity was a matter of simple blending of the traits of the parents, perhaps through some chemicals in the blood. For Darwin, life and its origin was just a matter of the same sort of chemical reactions in non-living nature. Although earlier researchers had demonstrated under a number of conditions that life was not generated from non-living matter, many biologists still believed that tiny, simple living things could form in nutritious material under the right conditions. Darwin privately speculated in a letter to a friend that perhaps some chemicals came together in a warm pool and became a living thing with the help of a natural electrical shock. Two of the greatest supporters of evolution at that time, T. H. Huxley and Ernst Haeckel, hoped to find some chemical blob still representing the border between almost-living and barely-alive somewhere, perhaps at the bottom of the sea. Some slime dredged up from the bottom of the ocean was actually thought to be just such an organism for a while, and before its true nature was discerned it was named *Bathybius haeckeli.*

Even in basic sciences not directly related to these, the world looked simpler and more intuitively natural then. When Darwin wrote and published his book, atoms were thought of as simple little solid balls, and light was simply a kind of wave, just like sound waves but in a much thinner material, called aether. The view of the universe on a large scale was also much simpler then. There was no knowledge of other galaxies, and, even more than in geology, nothing to contradict the belief that the universe had always existed just as it was. As far as the astronomers of the time could see, nothing was out there except ordinary stars, perhaps much larger or smaller or hotter or cooler or more variable than our Sun, but nothing truly extraordinary. Space had exactly three dimensions, and time was a totally separate thing that was the same everywhere for all potential observers.

[In what seems like more than mere coincidence, as I began reviewing this chapter, I also started to read Michael Crichton's *Timeline* (Copyright 1999 by Michael Crichton, published by The Ballantine Publishing Group). I was shocked to discover that the Introduction presents a very similar statement of the perception of the near-completeness of science in the late 19th century. Crichton backs up this assessment by quoting Alastair I.M. Rae, *Quantum Physics: Illusion or Reality?* (Cambridge, Eng.: Cambridge University Press, 1994), and adds: "See also Richard Feynman, *The Character of Physical Law* (Cambridge, Mass.: MIT PRess, 1965.) Also Rae, *Quantum Mechanics* (Bristol, Eng.: Hilger, 1986)." The last two references may be in support of another point he goes on to make which I also touch on in this book, that despite an ongoing or renewed faith that science is about to finish its grand work, there are indications that there are more surprises about to shake up established views. (Sadly, I later discovered that *Timeline* has some very bad language in it, so I cannot recommend it.)]

So in Darwin's day, it was easy for people to get the impression that science had discovered that the entire universe was like an infinite box of tiny but ordinary balls (atoms), just bouncing around in their ordinary way for an eternity, forming all possible combinations and shapes, until they happened to take the forms of all the material things we see now. Darwin merely supplied rationalizations for accepting the inevitable conclusion: that even life itself, including human life, was nothing more than the current stage in this endless but ordinary process. This attitude of scientific triumphalism makes it much more easy to understand why even strongly religious people sought ways, such as the Gap Theory and Theistic Evolution, to accommodate vast ages of prehistory and Darwinism.

Section II: Finally, Darwin.

Ideally, in scientific matters, there should be no need for considering personal information about a researcher. However, this is not an ideal world, and this book is an overview of historical and social as well as scientific matters. When creationists point out the flaws of Darwin's original work, many evolutionists will quickly deny that Darwin means all that much to them. On the other hand, when not put under such pressure, many evolutionists still seem to think it's important to praise Charles Darwin. Some actually celebrate Darwin Day! Darwinism seems a lot more glamorous and believable when it is portrayed as something that a genius was forced to believe, by sheer weight of scientific evidence, as a totally new idea, in the face of nearly-overwhelming religious opposition. Therefore, it will be worthwhile to look at Darwin as a scientist in his social context, and how he fits into his day and age, before examining his work itself.

Darwin is commonly portrayed as a towering genius, one of the greatest scientists of all time. He is also described as an original thinker, able to see what no one else could. Supposedly this made him a bold rebel against the traditions and common beliefs of his day. He's been made out to be something of a martyr for the cause of science, suffering the cruel slings and arrows of attacks by the religious establishment, until his great collection of scientific facts gradually won over all but his stubbornest and most ignorant detractors.

It should already be obvious that much of this can't be true. I've noted that evolutionary ideas had been put forth long before Darwin. Charles Darwin's own grandfather Erasmus had published a book with an evolutionary view. Lamarck's *Zoological Philosophy* had been published the same year that Darwin was born, and that wasn't the last of his evolutionary writings. Darwin noted, in an appendix to later editions of his *Origin*, a number of scientific researchers before him who had proposed similar theories, or key parts of his own. Although he left out several important 18th-century men who had also proposed evolutionary ideas in a less-scientific form, Darwin acknowledged he was not entirely original in his ideas about evolution.

It should also be clear that Darwin wasn't guided by some idealistic, unbiased search for the truth. Darwin was neither converted from a strong religious faith nor was he a complete and consistent Rationalist. He seems to have exhibited a blend of various major religious trends and philosophical ideas common in his time. His higher education had originally been intended to mold him into one of the liberal clergy that sons of well-to-do families often became, largely for the easy work and satisfying income. There's no indication that he ever felt a great zeal about religious matters. He was caught up in the growing popularity of science and was more interested in various lectures that introduced him to the latest ideas in geology and biology than in his theological studies. This, plus having connections with the right people, was enough to get him appointed as geologist on the exploratory voyage of the H.M.S. *Beagle*. The appointment may have been made largely to provide captain FitzRoy with company higher in social class than the common seamen on his ship.

Before Darwin had sailed in 1831, Washington Irving had already (1828) dreamed up a story about Columbus trying to justify his voyage of discovery, in which religious leaders argued against him on the grounds that the Bible taught that the Earth was flat, and anyone who sailed too far out would sail over the edge. Before Darwin returned to England in 1836, Antoine-Jean Letronne had contributed to this distorted view of the relationship between religion and science with his *On the Cosmographical Ideas of the Church Fathers* (1834). By the time Darwin published his ideas on evolution (December 1859), there were plenty of people ready to take up the cause of any modern, scientific idea that might replace a religious one.

The Church of England and some other religious groups had long before lost their religious fervor and replaced it with formalities and social concerns. Even some of the more traditional and fundamentalist churches had somewhat compromised their understanding of the Bible in order to accommodate uniformitarian geology. Darwin himself continued to attend church services (and was never expelled for his views), although he'd lost what little faith he'd had in God after his daughter Anne died in 1851. It seems he never became a complete atheist, but on the other hand there's no reliable information that he ever regained any faith in God. In later editions of his book, he added a concluding remark mentioning the "grandeur in this view of life... originally breathed by the Creator," but there's no indication that Darwin or any of his greatest supporters found the view truly uplifting or an incentive to believe in a Creator. There certainly were religious men who found evolutionism a supportive part of their modern faith.

This, then, is the true garden in which Darwinism grew. It had been around long before Darwin. Darwin was not highly trained in science, nor did he display a great originality of thought, any mathematical analysis, nor extensive experimentation in arguing for evolution. Rather than standing out ahead of his time, he was a perfect fit for the scientific, religious, and social world he lived in.

Darwin's view of life as a product of slow, natural processes had been shaped by Lyell's geology, which he studied while on the *Beagle*. Viewing the geological and biological data in this light as the voyage continued, Darwin gradually but inevitably acquired a vague vision of life also changing naturally, extremely slowly and gradually. There was no dramatic moment of discovery, not even when he was collecting varieties of "finches" on the Galapagos Islands. Meanwhile, this faith in Lyell's views on the unchanging laws and processes of nature, and time so long that no beginning was in sight, soon destroyed Darwin's belief in the Biblical view of God the Creator. Starting with ideas about variation and selection previously proposed, he slowly and privately sought ways to explain away the lack of evidence for evolution, and to respond to other difficulties.

He found the key to his biggest question in 1838, when he read a work on population growth by the Reverend Thomas Malthus (*Essay on the Principle of Population*, 1798). The fact that living things were capable of producing more descendants than the environment could sustain, and thus many would be forced to die without reproducing, appeared to be the means by which nature could select things with new traits the way ranchers select the best animals for breeding. A similar sentiment is found in Dickens' *A Christmas Carol* of 1844, in which Scrooge reacts to concerns that the poor may die by saying, "they had better hurry up and do it, and decrease the surplus population."

It's true that there were still a few Scriptural geologists publishing contrary views, and Darwin's own wife continued to believe in divine creation. These may not have been what kept Darwin from publishing, so much as his own doubts, feelings of unworthiness, and ill health, combined with his grand plans to compile a truly massive and definitive work in favor of evolution. At any rate, he continued to write in private and share his ideas with a few sympathetic friends such as Lyell, Joseph Hooker, and Thomas Huxley.

But Darwin was not alone in seeking a modern form of evolutionism. In 1852 the German philosopher Herbert Spencer published *The Development Hypothesis*, ("Development" is a translation of a German word that can also be translated "Evolution"). Darwin himself avoided using the term "evolution," preferring to speak of descent with modification. In 1855 Alfred Russel Wallace published an argument for evolution. In 1858 Wallace (who had also read Malthus' work) came down with a bad fever. In this state, he experienced what he felt was a flash of insight, revealing that natural selection was the means by which variation could lead to the evolution of new species. When he sent a paper on the subject to Darwin to get his opinion, Darwin recognized that Wallace's ideas were virtually identical to his own. He shared this revelation with his friends in the scientific establishment, Lyell and Joseph Hooker, who got together and arranged to have Wallace's paper presented to the scientific world – along with a letter Darwin had written earlier.

Darwin hurried to get his writings on evolution into publishable form, and just weeks before 1860 *The Origin of Species* was in the bookstores. Rather than being snubbed or burned by an overly-religious populace, it quickly sold out (although this may have been largely due to friends buying it in bulk quantities). At any rate, however much Darwin and his ideas were attacked and ridiculed, there's no record I know of that he ever suffered any significant social or religious ostracism, physical violence, or pressure from a scientific, governmental, or ecclesiastic body to change his views. On the contrary, there were supporters from theological circles as well as from a number of scientists, from the start. Darwinism was not so much a revolution as the capstone of the edifice of naturalistic philosophy.

Section III: Darwin's *Origin of the Species by the Preservation of Favoured Races*, its immediate aftermath and reception

Now that we've examined the social context of Darwin and his most famous book, let's look at *The Origin of the Species* itself.

One Essential Premise: Totally Naturalistic Philosophy

First of all, we must always keep in mind the underlying, essential, and yet almost totally unmentioned assumption at the heart of the book. Buried deep within the book is a statement that amounts to making a rule for a game that says "Only our team can play, the others lose by default": "On the ordinary view..., we can only say that so it is; that it has pleased the Creator to construct all the animals and plants in each great class on a uniform plan; but this is not a scientific explanation." (Chapter XIV, "Morphology" section, third paragraph).

The assumption is that science can account for everything, and can't acknowledge a limit beyond which other explanations may be needed. Once you've demanded an explanation that only allows natural processes, some form of evolution is the only remotely plausible answer. This is the major philosophical pillar that still supports evolutionism. Theistic evolutionists may believe in the supernatural realm, but they do not allow their god to do anything detectable in the universe that wouldn't have happened if there were no God.

Darwin had obviously inherited this philosophical naturalism from Lyell's uniformitarian geology. In the fully consistent application of this philosophy, there can be no miracles, and thus no valid non-scientific explanations; Scriptural revelation is made totally irrelevant to the physical world – it is not mentioned as such by Darwin even once. Although Lyell was slow to embrace his friend's notion of descent with modification in its fullest extent, it is clear that Darwin considered Lyell's uniformitarianism essential to acceptance of his own views. "He who can read Sir Charles Lyell's grand work on the Principles of Geology ... and yet does not admit how vast have been the past periods of time, may at once close this volume." (Darwin didn't share this caveat until Chapter X, in the first paragraph of the section "On the Lapse of Time, as Inferred from the Rate of Deposition and Extent of Denudation.")

The problem with this approach to science (inherited from deism and rationalism) is that it hides the lack of repeated observations and demonstration that are essential to the scientific method, and it limits or eliminates the possibility of falsification. "Science" of this sort is a matter of accepting whatever natural explanation seems most plausible. So it is not surprising that much of Darwin's book is aimed at clouding the deeper issues while covering up or attempting to explain away problems. It's easy to miss the fact that none of what he writes provides anything solid with which to decisively demonstrate his greater claims, nor does he allow that anything might prove an insurmountable difficulty for them.

One Stated Target: Fixity of Species

Another very important consideration for understanding Darwin's work is that it was not explicitly aimed at destroying belief in divine creation absolutely, nor was it aimed at creationism in general. At least, not on the surface. Darwin's stated main target (we might call it a straw-man or punching bag) was the doctrine of "fixity of species." This was the belief that the first ancestors of living organisms had been formed after the last of a long series of total extinctions. It also held that all species had remained exactly the same since that time. This notion would require that scientists had accurately identified species, and that no species had significantly degenerated, adapted to new conditions, or varied, let alone evolved into something more complex. A corollary was that species on islands had lived on them without changing since they had formed.

Darwin was thus able to use as counter arguments such trivial things as the breeding of pigeon varieties (as comparable to the formation of new species in the wild). This left the origin of larger groups largely untouched, and what he wrote about that was mostly vague variations of the "species are like domestic varieties" arguments, with hardly any consideration that evolution of all life from microbes might require something significantly different. Darwin sometimes wrote as if limiting his focus to within genera, but now and then indicated that all living things might have descended from a very few types, perhaps one simple form. As his main opponents also rejected the historicity of Genesis and largely accepted naturalism and indefinitely vast time, Darwin's theory was favored by the dominance of uniformitarianism and belief in absolutely inviolate natural law. Near the end of the book and in the appendix, in later editions, Darwin acknowledged that strict "fixity of species" was already becoming outdated by 1859.

Darwin's argument was also strengthened by the weakness of the concept of species. Darwin didn't try to define his understanding or use of the term "species," leaving it to a sort of common "we can't describe it but we know one when we see it" unwritten condition.

Perhaps if it were not for the confusion and mistakes in our classification of living organisms, people would have seen this argument by analogy as very weak. Darwin argued that domestic animals had been improved largely without conscious effort, and so the process was like the mindless forces of nature. However, even "primitive" breeding work had not been as "unconscious" as Darwin made out. Furthermore, the greatest results had been produced by very unnatural breeding programs, including artificially-controlled positive selection of mates and other factors not found in nature.

One Mechanism: Natural Selection

To make his idea of descent with modification seem new and workable, Darwin needed some natural process or phenomenon that would serve as a selecting agent, to complete the analogy with human's animal- and plant-breeding efforts. That mechanism was the Malthusian notion of an intense struggle for existence, which drastically eliminated all but the most fit. This desperate struggle, especially intense between the closest-related groups, supposedly provided a powerful, strict form of natural selection of organisms. According to Darwin, "the very slightest" advantage would tend to "favor" an organism's survival and reproduction. Natural selection was "not the exclusive means of modification," but Darwin considered "spontaneous variations" (recessive traits and mutations) a minor factor. In Darwin's own words, the heart of his theory is "... the doctrine of Malthus applied with manifold force to the whole animal and vegetable kingdoms..." (Chapter III, under "Geometrical Ratio of Increase"), "...one general law leading to the advancement of all organic beings – namely, multiply, vary, let the strongest live and weakest die." (Chapter VIII, summary)

One Driving Force: Boundless Variation

Natural selection alone, operating as it does by simply eliminating the less fit, could never produce any significant change. Besides, the idea of natural selection was not new, and was included in the creationist view of biology. As a consistent naturalistic philosopher, Darwin had to appeal to ordinary variations as a source of changes which could accumulate to produce not only new species, but far greater and different kinds of differences. Once again, the lack of knowledge at the time provided Darwin with an apparent strong point even though it was wrong, in that much of the variation Darwin thought was inheritable, isn't, and what is inheritable isn't really new, but comes from combinations of ancestral genes. Most importantly, he was wrong about ordinary variation being driven slowly (over many generations) by environmental conditions, habits, and use or disuse of parts (e.g. larger udders on goats and cows due to constant milking for generations).

Even as Darwin was working on his book, there were a number of observations that undercut his notion of boundless variations accumulating with the elimination of less-fit variations. He lightly dismissed what indications there were at the time of limits to variation by artificial selection. Darwin also downplayed the major problem of the tendency in domestic species of reversion towards the common, wild conditions. In the natural order of things, generational variations are limited, and only a vast Intelligence could have produced the panoply of all the different amazingly complex living things.

Major Supporting Observation: Minor Variations

Besides pointing to the variations in domestic plants and animals, Darwin pointed to a number of observations of variations in the wild. The most famous example is the different "species" of "finches" in the Galapagos Islands. Although the differences between them are not very large, examples such as that were more than enough to knock down the tottering doctrine of Fixity of Species.

Another observation that Darwin leaned on, as do his followers today, is the similarities between various organisms. The reasoning is that, if differences show that living things can vary indefinitely, then similarities must show that they came from the same ancestor. In a science based on telling stories about the past, opposite things can support the same theory. Not only is any other explanation ruled out as unscientific, but everything can be turned into support for the idea.

Darwin also argued that there were observations which, if only indirectly, showed that major transitions had occurred. While paleontology was still an infant science, most of the fossils that had been found were either quite similar to living forms, or entirely new forms that obviously were not transitional and thus would require new transitional lines of their own. Most of Darwin's writing on the subject was therefore attempting to explain away the vast gulf between what his theory predicted and what was actually observed. However, he was again greatly aided by the easy target he had chosen to directly address, and by the philosophical monopoly of the science involved.

Darwin did not, in this new way of doing science, have to demonstrate that any major transitions had happened, or even that they were possible, only that they were the most plausible natural explanation for the origin of things as they are now. Thus, any form of life that showed some combination of similarities and differences that are shared by two other forms can be said to illustrate a step in the line of actual transitional forms. Such explanations sound more impressive if you don't consider that a three-wheeled motorcycle could be said to illustrate a transitional stage in the inventive progress from a bicycle to a car. We should also keep in mind that the progressive spirit of the age favored the notion that living fish, amphibians, reptiles, and mammals might illustrate a history of progress from the ocean depths through the murky swamps and so forth, "onward and upward."

While Darwin imagined complete lines of populations slowly changing, he also dreamed up stories as to why transitional forms might never be found as fossils. One was that all the major transitions might have taken place in relatively small areas that were not prone to producing fossils. This was more of a rhetorical device than a scientific hypothesis, but it did provide an explanation for the lack of evidence. It had the additional advantage of making any possibly transitional fossil seem to be a major support for evolution. Not long after Darwin's book was published, the popular hunger for exciting "missing links" was ignited by the discovery of Archaeopteryx. This excitement has been fed by one or two widely-publicized discoveries every now and then. Further studies showing the animals were more highly-specialized and less like transitional forms tend to be left out of popular media. Thus, while fossils overwhelmingly show distinct forms, not evolution, this fact is lost in the propaganda about a few "intermediate" forms that supposedly represent all the missing multitudes of forms, and so "fill the gaps" with imagined transitions.

Proposed Tests: None.

As far as the notion of Fixity of Species was concerned, Darwin didn't need to do much more than point to some cases of mistaken species identity, and compare the known variety in domestic animals with the narrowly-defined species of wild animals. On the other hand, Darwin also ventured into speculation about the origin of larger classifications of living things, including the development of the most complex forms of life from simpler ones. He admitted, however, that the evidence for this was weak and might never be strong. It was only after another decade, when his theory was largely accepted, that he dared to explicitly propose that even humans were descended from simpler forms.

Darwin came close to proposing a test of the greater implications of the theory, by hinting that perhaps the fossil record would eventually support his idea of continuous lineages linking all life forms to a very few common ancestors (or just one). However, he did not try to estimate how long it might take, or the number of fossils required. Sadly, his further suggestion that fossilization might be biased against preserving transitional organisms backed away from the earlier hint that the fossil record might be a test – at least, not if it didn't support the theory. Again, Darwin framed the possibilities so that any evidence that was merely consistent with his theory was taken as good support, while the lack of any amount of expected evidence was supposedly not a problem.

Darwin did predict (near the end of the last chapter) some effects that acceptance of his theory would have on science, but these predictions were self-fulfilling, circular (for example, if the theory was accepted, then data would be evaluated in light of the theory), or compatible with creationism. For example, he predicted that classification would be changed and made more certain, but today's "evolutionary" practice of classification doesn't depend on the theory of evolution, although the results are interpreted in evolutionary terms. The binomial system of naming within a hierarchical order of classification is still based on the one created by Linnaeus, a creationist. It's not a perfect system, but at any rate the notion of evolutionary relationships hasn't cleared up the difficulties of accurately classifying everything.

Darwin also made a statement which, if read lightly and out of context, appears to offer a means of falsifying his notion. "If it could be demonstrated that any complex organ existed, which could not possibly have been formed by numerous, successive, slight modifications, my theory would absolutely break down." However, rather than being a true test of the theory, it is merely a rhetorical trick allowing Darwin and his followers to shirk responsibility for the burden of proof. The proponent of a new theory should demonstrate that what is proposed is at least possible, even when it must be left to others to test if it does or did occur. It is far easier to prove that something is at least possible, than to prove that there isn't any possibility – especially when the only true alternative has been ruled out of discussion in the first place!

In practice this approach allowed Darwin to substitute his imaginative stories in place of the precise and repeatable demonstrations that have always been the hallmark of science. Today, the exceptions to the scientific method are still in fields related to impractical beliefs about an unobserved past and evolutionism, where researchers need only suggest plausible stories, and then sit back and see if anyone else cares to suggest something else – and only perfectly naturalistic explanations are allowed. Like Darwin they demand that doubters provide all-encompassing disproofs with every possibility accounted for, but examples of complexity that can't be explained by "numerous, successive, slight modifications" are "refuted" in their minds by their stories that paste together scattered bits of data with imagination, and faith that it had to have happened naturally somehow.

One Fundamental Logical Flaw

The fatal flaw of Darwinism is in its fundamental premise that life and all its various wonderful forms must be accounted for by slow accumulation of small, natural variations, with no intelligent guidance. It not only creates prejudicial censorship, it includes the false assumption that changes differ only in degree. Since Lyell's uniformitarian geology had already been firmly established in scientific circles, it was a big boost to acceptance of Darwin's notion that any kind of change could be simply a matter of accumulation of small changes, like the slow rise of mountains. But when applied to living things, this is exactly equivalent to reasoning that since wind can blow sand into large dunes, natural processes alone can also turn the sand into computer chips!

This is not simply an extrapolation (assuming something could continue in the same way beyond what has been observed), but an unacknowledged leap from one category to another. To give an ordinary example, painting a bike does not change it in the same way as adding training wheels. For that matter, neither does taking off the training wheels, or whacking the bike with a hammer in random places. No matter how much you paint it or hammer it, you won't add complex moving parts to it. In living things, changes in colors, loss of parts, or parts becoming longer, shorter or somewhat differently shaped, might change something to the point that scientists would move it into a different species or genus, but no number, degree or combination of such changes can increase the systematically organized dynamic complexity and produce an entirely new kind of organism. Without such increases, descendants of a simple organism could never become any of the more complex forms.

Rhetorical Techniques

While Darwin's masterpiece leaves much to be desired as a scientific work, it is a *tour de force* of arguing tricks. Many of them are perfectly valid (I use them, too), but while attractive argumentation is a key feature of philosophy, debate, and legal representation, it is only window dressing in (most areas of) science. When used to distract people so they don't notice (or excuse) a lack of observation and demonstration, it dishonestly dresses up fairy tales to look like science. Perhaps Darwin had learned a few tricks from Lyell, who had been trained as a lawyer and used similar rhetoric to support uniformitarian geology. Darwin also demonstrated talent as a storyteller, invoking hypothetical situations and imaginary past events in ways that made them seem like facts.

PROBLEM SANDWICH: Darwin organized his book like a sandwich: First he presented a simplified version of his idea (skipping over problems and complicating factors), then presented (but dismissed) objections and problems, and finished with what he regarded as the most persuasive evidences in favor of his theory. In this way, Darwin was able to give the impression of fairly presenting his theory while in reality giving it every advantage: hiding the difficulties between the good stuff, and not giving due consideration to any other possibility.

MISDIRECTION (with Straw-man): Darwin's main ploy was to get the reader to focus on the problems with fixity of the species, when the real problem was explaining the origin of all the distinctly different forms of life. He also used this technique on specific points. For example: Darwin suggested the reasonable supposition that a pigeon breeder couldn't imagine breeding a fantail pigeon until a mutation produced a pigeon with extra tail feathers, and then the more doubtful supposition that the breeder could not imagine what later fantails would look like.

This supported Darwin's notion of "unconscious" selection (which made artificial selection seem more like natural selection), because it distracts readers from the question of how useful new changes are supposed to arise in nature, and draws them into Darwin's imaginary world where breeders would not use their brains in recognizing any changes, controlling the breeding in the population, imagining that some further increase in the number of tail feathers might occur, and ensuring the survival of the breed even if the changes make individuals less likely to survive and reproduce.

Misdirection is a fundamental principle of the art of stage magic. The stage magician's use of smoke and mirrors has become a metaphor for purposely creating deceptive appearances. Darwin used a number of techniques which can be described as rhetorical equivalents of smoke and mirrors.

SMOKE techniques (AKA Jelly, as in the saying "it's like trying to nail jelly to the wall"): Darwin often made vague and unsupported statements of the "probability" of things supporting the theory – and based further points on these statements. For example, he used imaginative suppositions and supposedly reasonable stories about giraffe necks as support, rather than controlled experiments with quantitative data. He explicitly asked the reader to be satisfied with nothing more, e.g. "I must beg permission to give one or two imaginary illustrations." While Darwin often invited the reader to think along with him, he did not trouble himself to think through all the details involved, or to give due consideration to other possibilities. For instance, he did not consider how sexual dimorphism (significant differences between males and females) might negatively affect his imagined balance of variation and selection for survival of the fittest.

Sometimes Darwin's vagueness seems to be due to confusion or indecision. For example, compare his discussion of "nascent" vs "rudimentary" or "reduced" (i.e. vestigial) parts with his musings on parts being reduced although of "high service" (e.g. penguin flippers as vestigial wings); and with his struggle with the question of how vestigial parts might be completely eliminated, although there would be very little selective effect once they were highly reduced. Whether an intentional ploy or a lack of mental depth and clarity, this vagueness and flip-flopping can cloud the unwary reader's thoughts with a sense of sympathy and plausibility. (see Chapter XIV: Mutual Affinities of Organic Beings: Morphology – Embryology – Rudimentary Organs) Here again, the reader is directed to focus on these as evidence for common ancestry, although in these cases the descendants have reduced or lost features, just the opposite of what needs to happen to go from microbes to all other forms of life.

Hedging: Darwin constantly used vague, softening modifiers such as "generally," "tends to," etc. Counter-examples then could be explained away as exceptional. Darwin sometimes went as far as to provide explicit logical escape hatches, so that hardly anything he said, or any conceivable new data, could be used to pin him down. Likewise, the hypothetical concept of "ghost lineages," a technique popular today, provides an escape from the fact of fossils being dated in the opposite order from the evolutionary series they are said to illustrate.

Daffynitions: Darwin had the advantage of existing vague and mistaken ideas about species, and clearly made the most of this and any other helpful blurring of definitions he could use. For example: "the best definition which has ever been given of a high standard of organization, is the degree to which the parts have been specialized or differentiated..." There's a fundamental difference between "highly organized" and "highly specialized." Increasing specialization and differentiation can be accomplished by simple modifications of existing parts, but increased organization only comes with the addition of new parts. This is one of the favorite techniques of modern evolutionists, who like to confuse "organization" with "order," a much simpler condition.

MIRROR techniques: The prime example of this technique, from Darwin to the present, is describing the negative filtering effect of natural selection (weeding out the unfit) with mirror-image positive terms such as "preservation" of traits and "favored" races. However, there are several other rhetorical tricks that can be compared to using mirrors to create false images.

New Deities: Darwin created a mirror-image substitute for the power of intelligence, under a thin guise of metaphor, in attributing to natural selection or nature various anthropomorphic functions such as "watching for" favorable variations, "working for the good of each being," and "taking advantage of all favorable variations." This pseudo-intelligent "behavior" of natural forces is explicitly and favorably compared to the creation of works of human intellect such as magnificent architecture and telescopes. Time and Chance are also substitute deities which evolutionists have called upon to help Natural Selection and Nature accomplish unobserved miracles. (As Darwin said of the leaps of faith he saw in a rival evolutionary theory, "To admit all this is, as it seems to me, to enter into the realms of miracle, and to leave those of science.") Sometimes organisms are said to have developed innovations, devised contrivances, etc., as if they were PhD geneticists that experimented on themselves.

Funhouse Mirrors: Darwin used (and lightly asked to be excused for it) metaphorical and anthropomorphic terms for other purposes. Even "natural selection" is an example of this. "In the literal sense of the word, no doubt, natural selection is a false term... such metaphorical expressions... are almost necessary for brevity. So again it is difficult to avoid personifying the word Nature; but I mean by nature, only the aggregate action and product of many natural laws, and by laws the sequence of events..." (Chapter IV: Natural Selection; or the Survival of the Fittest, second paragraph) This distorts the reader's mental picture of the theory, as this shorthand language suggests the activity of an intelligent being, as we experience every day, making the theory seem more familiar and plausible, and less likely to invite close examination.

SMOKE & MIRRORS: Several of Darwin's rhetorical traps seem to involve both clouding an issue and presenting a false image of one or more parts of it.

Primrose Lane: This is the practice of starting with the fog-spreading way of stating things provisionally, as suppositions (with friendly, reassuring phrases such as "we may feel sure that..."), then making the statements more clear and positive until (as if in a mirror universe where suppositions are reality) stating the same things as plain facts. For example: In the section on "Unconscious Selection" (chapter I), Darwin referred to ancient animal breeding as "a form of selection, which may be called unconscious," then he shortened this to "almost unconscious," and later he wrote "partly unconscious and partly methodical, selection." Finally, he used the phrase "unconscious means of selection" as if it were settled that our ancestors, who invented domestication and selective breeding, were mindless drudges.

Bait-and-Switch: Focus on the straw-man (smoke), then toss in or wave at (mirror) the real question without properly dealing with it. This was Darwin's overall approach, attacking the notion that living things hardly changed over generations at all, and using a lot of imagination to create a mental image of a "tree of life," but never demonstrating that universal common descent was possible. It is still a popular technique of evolutionists, who often probably don't even realize they are using it.

Rubber & Glue: (As children used to taunt, "I'm rubber and you're glue...") One of Darwin's favorite techniques. His merely imagined possibilities (smoke) were used as support to downplay any problem as being "not insuperable," while (mirror) he claimed that problems with other views were without any imaginable resolution. He suggested that what was unknown should lead his opponents to be humbly cautious, but he always supposed that when previously unknown things would be learned, they would favor his view. For example: "Some authors have maintained that the amount of variation... is soon reached, and can never afterward be exceeded. It would be somewhat rash to assert that the limit has been attained... for almost all {the greatest improvements in livestock had been accomplished} within a recent period... It would be equally rash to assert that characters now increased to their utmost limit, could not, after remaining fixed for many centuries, again vary under new conditions of life." (Chapter I, third paragraph under "Circumstances favorable to man's power of selection") For another example, he argued that the gaps in the fossil record shouldn't be used against the theory, but felt it safe to suppose that if any fossils did turn up, they would be supportive, without considering that entirely new forms might be discovered, requiring yet more undiscovered transitional forms. He argued that the nested pattern of taxonomy (classifying animals in groups within groups) is perfectly explained by his theory, but there couldn't possibly be any explanation if species were created – although the system had been created and developed by creationists who had perfectly logical reasons for seeing it as evidence for creation!

"JUST SO" STORIES: Creationists have compared the practice of evolutionism to telling "just so" stories, a reference to Rudyard Kipling's children's stories in which fanciful explanations were presented as the ways that things turned out "just so." Darwin used several variations of this technique.

Goldilocks, or Pollyanna: Supposing that everything would turn out "just right;" having blinders on, tunnel vision, or selective memory regarding anything contrary. In one place, Darwin discussed the advantages (in his mind) of different population sizes on variation, and in another place he described the effects of different population sizes on extinction rates, producing a positive if dream-like sense of support for his theory. Throughout all this, he always envisioned every effect as helping his theory. He never considered the probability that different effects might counteract each other, or anything like that. He regarded the extinctions of parent and variant types as great enough to leave clear distinctions between old and new, and yet also not so much as to prevent older types from continuing to exist. He saw an evolutionary tendency towards perfection in nature that is pervasive and powerful enough to produce all the wonderful adaptations and complexity in living things, and yet at the same time just slow and ineffective enough to explain any (real or apparent) imperfections or relative simplicity of any organisms. In Darwin's opinion, the conditions for fossilization were just common enough to preserve a few arguably intermediate forms, and yet bad enough to explain away the absence of the predicted vast numbers of continuous lineages.

This way of looking at things is still used to explain both the lack of change in many organisms from ancient fossil forms to living examples, and (at the same time) changes in others being far more rapid than usual. Darwin at one point wrote that natural selection can catch changes finer than the variations of a few feathers that pigeon breeders have built on, but in another place he wrote that "It cannot have been of much importance to the greater number of mammals, birds, or reptiles, whether they were clothed with hair, feathers, or scales..." Whatever evidence there is, is always described as "just right" for evolution, a habit which leads to the next technique.

Procrustean Bed: an extreme form of the Goldilocks technique, in which any datum, no matter how different from other data or contradictory to previous evolutionary explanations, is made to fit the theory. This was built into the 18th-century vision of science as always providing only perfectly natural, purpose-free explanations for everything. To Darwin, the power of variation and selection explained the origin of the marvelous giraffe, but he also "saw no difficulty" in the absence of any other living mammals with such long necks, nor in the many short-necked animals living alongside the giraffes. The theory "explains" why a number of land animals supposedly "went back to" living in the sea, yet supposedly there's no problem in fish supposedly having produced descendants fully adapted to land once, but afterwards no fully aquatic animal ever again evolving the ability to live fully on land. Darwin appealed to observations of slight tail-wrapping in African monkeys to illustrate his imaginary tale of the origin of prehensile tails in New World monkeys, but expressed no qualms at the fact that none of the African monkeys had evolved prehensile tails.

Silence is Golden: Simply ignoring important points entirely if too much trouble. For example, Darwin compares the weeding out of the unfit (leaving only the most fit) with human breeding of the best-quality stock, but never addresses the fact that the practice of artificial selection will often allow only one male in a population to act as stud, and breeders will also transport stud animals from one place (and herd of females) to another, far beyond the range natural behavior would take them.

See No Evil, Speak No Evil: The Rubber half of the Rubber & Glue technique, used alone or with "Silence is Golden." For example, Darwin would state that something "offers no difficulty" or "I can see no difficulty..." or "there is no difficulty in supposing..." to imagine whatever data was needed to support the theory. To cover up the lack of serious consideration of major drawbacks, weaknesses, and contrary observations, such statements were instead typically followed up by more rosy clouds of suppositional smoke, hedging the bet, etc. ("On our theory the continued existence of lowly organisms offers no difficulty; for natural selection, or survival of the fittest, does not necessarily include progressive development – it only takes advantage of such variations as arise and are beneficial..." – under "On the Degree to which Organization Tends to Advance" in Chapter IV) In at least one case, a serious gap was acknowledged, but the seriousness of the problem it presented was baldly shrugged off: "Mr. Busk, however, does not know of any gradations now existing between a zooid and an avicularium. It is therefore impossible to conjecture by what serviceable gradations the one could have been converted into the other, but it by no means follows from this that such gradations have not existed." (Chapter VII)

Buddy-buddy talk: Darwin used inclusive pronouns ("our theory," "we can see") at various points that the reader might well not agree with, a familiarity and appeal to fellowship and camaraderie designed to dull any healthy scientific skepticism.

"Mounds of evidence": Giving a number of examples (and hand-waving statements about there being many more) which are merely variations of the same thing and don't address the larger questions or harder targets. This waving at the sheer amount of data covers up the fact that there are no experiments to show various things could actually happen.

Hand Grenades and Horseshoes: For Darwin, "close enough" was good enough – when it helped. Whatever problem confronted his theory, if he could vaguely imagine an explanation that made him feel that "the difficulty is not insuperable in believing" his theory was still sound, it was as good as if he'd demonstrated that the problem was solved. Very effective when combined with Rubber & Glue and a reader lulled into complacency.

Appeal to Authority: This was a return to an outdated practice of Medieval Scholasticism, which scientific researchers up to this point had rejected as the main problem with philosophy and education in general. However, as Darwin was unable to demonstrate the possibility of key aspects of his theory, he substituted appeal to the opinion of scientific experts whenever he could. Sadly, appeal to the authority of consensus and expert opinion is still one of the major pillars of evolutionism.

Repetition: Whether through a naturally plodding habit of thought or as an intentional persuasive device, Darwin repeatedly used favored terminology of his theory, for example: "slightest variations," "extremely slight modifications," "slightest difference," etc.

Playing Don Quixote: In the chapter (VI) on "Difficulties of the Theory," Darwin introduces the subject with what appears to be a shocking admission, "...I can hardly reflect on them without being in some degree staggered." He writes similar things about several individual cases. Several problems were very challenging, and remain major problems for the theory, but it seems Darwin's admissions were largely a rhetorical technique, as each one was followed soon after by more of his special imaginative suppositions, the Primrose Lane, and other techniques. By stating the appearance of the problems at first in strong, even emotional terms, Darwin's flimsy stories, with their superficial "solutions" to the problems, followed up by his Pollyanna-style confidence, could appear to have turned giants into windmills, and the need for a real resolution might seem to vanish. Viewed with a critical eye, Darwin's statement, "...to the best of my judgment, the greater number are only apparent, and those that are real are not, I think, fatal to the theory" is obviously a weak statement of personal faith, but to the unwary reader impressed by Darwin's apparent recovery from a "staggering" blow, it might well sound like a statement of decisive scientific victory.

Reversing the Film: In a famous bit of WWII Allied propaganda, a film of Hitler walking was edited to show him going forward and back, as if doing a little jig of evil joy. Darwin pointed to different degrees of loss of function (for example, apparent loss of flight in birds) and suggested that "they serve to show what diversified means of transition are at least possible." (Chapter VI, under "On the Origin and Transition of Organic Beings with Peculiar Habits and Structure") This argument ignores the fact that it is always much easier for a complex function or structure to be lost than to be formed. A child with a hammer can easily ruin many electronic devices, but you can't point to various stages of the destruction and say it illustrates how a device could form.

Saving Tinkerbell: That evolutionism is based on faith, not objective science, is seen by the number of times Darwin felt compelled to note, "I believe..." or suggest that "we may believe...", etc. At one point he wrote, "I must confess, that, with all my faith in natural selection..." (Chapter 8, the paragraph before the Summary) In a children's play, enough people clapping or shouting "I believe in fairies" is enough to revive Tinkerbell, but appealing to belief or the number of scientists who believe something is not the way to maintain a scientific theory.

While little remains of Darwin's theory in current evolutionary thought, the main ideas are unavoidable in any philosophically naturalistic system of belief. In any such scheme, the sudden appearance of complex forms is inadmissible. Therefore all living things must have come from simpler forms, by a gradual accumulation of small, naturally-caused changes. Since there is no natural force that can act positively to select for complex traits, the accumulation can at most only add a number of similar simple variations. Since no one has ever observed a purely natural process produce an increase in organized, dynamic complexity in living things (or anything else) that wasn't carried over from a previous generation, true believers in the system must have faith in some extremely gradual and/or rare mechanism. That is the true legacy of Darwin: a system of faith hiding under the trappings of scientific theory and data, which cannot be disproved within the system, and which will not tolerate criticism or rivalry from without.

Besides giving the notion of evolution respectability and giving model examples of many rhetorical devices for his followers to use, Darwin mentioned or made room for some specific concepts which are still popular. He used the term "coadpatation," and presented the idea of pre-adaptations, that is, changes which were useful for one function, but could also serve another function when other changes occurred later. He even allowed some room for stasis (lack of change) and punctuated equilibria (in its current, moderate form, not the "hopeful monster" version). Another concept he hinted at which later became important to evolutionists is neutral mutations: "I am inclined to suspect that we see, at least in some of these polymorphic genera, variations which are of no service or disservice... and... have not been seized on and rendered definite by natural selection..." (Did you catch the Jelly and Goldilocks aspects of this statement?)

It's no wonder that evolutionists, who still use such techniques, admire Darwin so much.

Darwin's Place Today

So, Darwin's work wasn't ahead of its time, it didn't directly attack any religious doctrine, and it didn't provide any scientific demonstration or overwhelming body of evidence contrary to a religious doctrine. On the face of it, it was only a dull technical essay from a man of impeccable British social class who had written a popular book about his voyage of scientific exploration on the *Beagle* a few decades earlier.

So, why all the fuss? It provided an acceptable biological application of the same philosophy of natural uniformitarianism as the Hutton-Lyell geology. Liberal clergy as well as scientists and philosophers with atheistic leanings could see much that favored their views in the theory, although these further implications were not spelled out.

So Darwin became a symbol, a figurehead for all who rejected traditional religious beliefs, and his theory the justification for new beliefs. In the book *Evolution and Other Fairy Tales* by Larry Azar, Dr. Azar quotes Ernst Mayr (arguably the most prominent evolutionist of the 20th century) on the revolutionary nature of Darwinism: "It is now clear why the Darwinian revolution is so different from all other scientific revolutions. It required not merely the replacement of one scientific theory by another, but, in fact, the rejection of at least six widely held basic beliefs..." The beliefs he listed were: the age of the Earth, catastrophism and a steady-state world, the concept of an automatic upward evolution, creationism, essentialism and nominalism, and anthropocentrism. (Chapter 4, "The Reception of Darwinism," p. 47)

This may give Darwin a little more than his due, because the age of the Earth, catastrophism, and Biblical creationism had already been largely rejected by many geologists, liberal theologians, and their followers. But Darwinism did bring all these points to the attention of the general public, and began to make them popular. They were shocking enough to cause some doubts and concerns among Darwin's supporters. Particularly galling to the more theistic uniformitarians was the implicit rejection of any active plan, purpose, progress or higher organizing principle. Darwinism argues that all living things are nothing more than the result of a mindless process – even the expression "struggle to survive" is a metaphor that implies more intent and purposeful effort than the scientific concept allows. While many people are happy with the mental gymnastics of believing that "God created through evolution," both great evangelists and leading evolutionists have seen that evolutionism is not compatible with any form of active theism. More than that, people from both sides of the debate have seen that the evolutionism, taken to its full logical consequences, undermines all sorts of social values and beliefs.

H.G. Wells, equating natural selection with Darwinism, wrote that as it "has been more and more thoroughly assimilated and understood by the general mind, it has destroyed, quietly but entirely, the belief in human equality which is implicit in all the 'Liberalizing' movements of the world – it has become apparent that whole masses of human population are, as a whole, inferior in their claim upon the future to other masses, that they cannot be given other opportunities or trusted with power as the superior peoples are trusted, that their characteristic weaknesses are contagious and detrimental in the civilizing fabric, and that their range of incapacities tempts and demoralizes the strong. To give them equality is to sink to their level, to protect and cherish them is to be swamped in their fecundity."[1] Furthermore, Wells concluded that "Darwinism destroyed the dogma of the Fall upon which the whole intellectual fabric of Christianity rests. For without a Fall there is no redemption, and the whole theory and meaning of the Pauline system is vain."[2] [1 & 2 Wells, H.G., *Anticipations of the Reactions of Mechanical and Scientific Progress Upon Human Life and Thought*, Bernhard Tauchnitz, Leipzig, 1902; reprinted by Dover, Mineola, NY, pp. 162-163, 1999, as quoted by Jerry Bergman in his article, "H.G. Wells: Darwin's disciple and eugenicist extraordinaire," *TJ* 18(3):116-120, December 2004]

Section IV: The Triumph of Darwinism

While some people were able to see the implications of Darwinism that were contrary to Christianity and societal ideas such as the fundamental equality of all men, the essential superiority of humans over animals, and the implicit human stewardship of the Earth, there were a number of factors in place at the time which in hindsight make the widespread acceptance and establishment of Darwinism easily understood. Above all else, as noted from the beginning, the introduction of naturalistic philosophy as an integral part of science made something like Darwinism inevitable and unassailable within the new scientific establishment.

However, there were several other factors setting up Darwin as the figurehead of the new evolutionary faith. Darwin had a very respectable but not aristocratic social standing, at a time when such things were still very important. The time was ripe, as seen by the publication of similar ideas, especially Wallace's letter which forced

Darwin to publish before he felt ready. Then there was Darwin's masterfully imaginative use of storytelling and rhetoric, showing his audience the unquestionable minor changes in domesticated animals and subtly making it seem quite logical to believe that all the different kinds of life could be produced from microbes by nothing more than continuing such changes, if given enough time.

But still, Darwin was just a figurehead from the start. He provided no precise mathematical calculations or useful, distinctive predictions, no new experimental demonstrations that might be repeated or exposed as flawed or fakes. As in geology, the standards and limits of science had been cast aside, and the door opened to young biologists eager to join the new, exciting game of creating a vast new history of the Earth by telling stories connecting points of data. It was in tune with the spirit of the age to feel confident that such "scientific" tales were true revelations about the world.

There were a number of contemporaries of Darwin who might each warrant a book of their own, just for their relationship to Darwin and his vision. While each of them seems to have had some doubts or reservations about some aspect or other of Darwin's theory or its implications, they supported it strongly as a whole. Charles Lyell and Joseph Hooker were the leaders of those in the old guard who embraced Darwinism, while Thomas Huxley and Ernst Haeckel became the greatest proponents and popularizers of Evolutionism for years to come.

It may be that the slight differences between Darwin's evolutionism and the variations presented by his supporters only helped along the acceptance of his malleable, open-ended concept. All sorts of views can fit under the big tent of variations selected by natural conditions. Theistic evolutionists can accept everything about it and still believe that God had something to do with it. On the other hand, even some atheists have had difficulty accepting the complete lack of purpose, progress, and intelligence. Darwin's nebulous theory provided room for its supporters to pick and choose on the specifics, and create all sorts of variations. Thus, any criticism of any specific point might be deflected by arguing that some other variety of Evolutionism held the key – and Darwin would remain as the symbolic father of it.

An illustration of a specific variation which still supported Evolutionism in general can be found in a review of a book about Haeckel's illustrations ("Painting the whole picture?" *NATURE* 445, 1 Feb. 2007, pp. 486-487, reviewing *Visions of Nature: The Art and Science of Ernst Haeckel* by Olaf Breidbach). The review points out Haeckel's use of overly simplified and unnaturally perfect drawings when he could have used microphotography. It notes that Haeckel did produce a book of microphotographs, *Nature as an Artist*, "a series of photographs of his subjects that demonstrates, he said, that 'there can be no talk of reconstruction, touching up, schematization or indeed forgery' in his drawings." The reviewer (Phillip Ball, "a consultant editor for *NATURE*") doesn't say so at this point, but from earlier and later comments it's clear he thought Haeckel's photo book didn't fulfill its purpose. He says early on that Haeckel was motivated by a desire to twist Darwin's random ("pragmatic rule of contingency") scheme so it would fit Hegel's philosophy, producing "a search for order, systematization and hierarchy that would reveal far more logic and purpose in life than a mere struggle for survival." That's where Haeckel's "biogenetic law" comes in. Ball later states that "Haeckel's extraordinary drawings were not made to support his arguments about evolution and morphogenesis; rather, they actually were the arguments." Haeckel wanted people to see that what he believed was obvious just from looking – but as embryologists have attested since, there are differences in embryos that don't fit Haeckel's belief.

As far as support from social groups is concerned, the biggest boost to the success of Darwinism was probably the widespread and deep liberalization of the Church of England and other ecclesiastic organizations. Deists and other nominally Christian theistic philosophers who opposed traditional, Biblical Christianity had been around since the mid-1600s or so. By the mid-1700s, religion was openly mocked by most of the upper classes. By the time Darwin published his *Origin,* many leading clergymen and theologians were prepared to abandon any doctrine spoken against by philosophers and scientists. Darwin's supporters included the prominent theologians Frederick Dennison Maurice and Aubrey Moore. Perhaps the most telling single incident of the time happened in 1860, just months after Darwin's book came out: the publication of *Essays and Reviews* by "seven liberal essayists, six of whom were clergymen in the Anglican Church." (*The Faces of Origins: A Historical Survey of the Underlying Assumptions from the Early Church to Postmodernism,* David Herbert, M.A., M. Div., Ed. D.; D & I Herbert Publishing, London, Ontario, 2004) The essays showed these men in positions of religious leadership to be "avowed unbelievers in revealed religion." They explicitly elevated human reason over the Bible.

As Darwin's followers quickly gained the upper hand, they also swiftly moved to secure their position by portraying it as the only reasonable one, portraying any opposition as willfully ignorant and superstitiously opposing science and the progressive spirit of the age in general. They took as examples a biased account of Galileo's "persecution" and the fictional works of Irving and Letronne, and built on them. The geologists had already demonstrated how effective censorship by leaders of scientific organizations could be, by suppressing the Scriptural Geologists' works.

The Darwinists were so successful that the monopoly and censorship within major journals, academia, and major media is almost absolute. You may seldom read anything or meet anyone who knows anything about the reception of Darwin's book beyond the popular myth about ignorant church leaders denouncing Darwin while all the scientists were being convinced by the evidence. You will probably not hear nor read anything about the famous debate between "Darwin's bulldog" Thomas Huxley and Bishop Wilberforce except for the apocryphal exchange at the end, in which Huxley got the last and best word.

Two very powerful propaganda tools of the evolutionary establishment appeared before the end of the 19th century. The first was John William Draper's *History of the Conflict between Religion and Science,* in 1874, and the second was Andrew Dixon White's *A History of the Warfare of Science with Theology in Christendom* in 1896. (As noted in *The Politically Incorrect Guide to Science* by Tom Bethell, pp. 184-185) These and similar works spread through the education system and popular culture the simplistic myth that scientists were all objective rationalists, and they were constantly being opposed by religious leaders who knew nothing of science and saw it only as a threat to their power.

This indoctrination was so successful that by 1925 H.L. Mencken and other reporters had a large, appreciative audience for their cynical views of the Scopes Trial, portraying everyone in favor of the law as ignorant hicks and religious fanatics. When J. Lawrence and R.E. Lee were looking to write a play in 1954 as an allegory against the excesses of anti-Communism efforts, they produced *Inherit the Wind*. This melodrama is clearly based on the Scopes trial, but the people opposed to evolutionism are portrayed as so evil or ignorant that the story is a dark fantasy. Sadly, this grossly distorted image is all that most people are likely to have of the trial. When the facts of the trial do happen to be presented, it is not unusual to see references to the scientific evidence which a number of prominent scientists wished to have read into the trial records, but no mention that much of it is now known to be false. Popular accounts might refer to students testifying that they felt no harm from being taught evolution, but they are not likely to mention that the textbook in question espoused the sort of racist ideas and eugenics which led to a great deal of harm. Look up and watch the movie *Alleged*, which gives a more realistic view of the trial (admittedly solidly on the other side of the issue and with other imperfections, but a good contrast to the more famous movie).

As evolutionism became more and more established, it also developed a tendency to become hidebound, inflexible; so that even variations within the framework of naturalistic philosophy met with opposition and had to fight for acceptance. Unlike other fields of science, evolutionism still has never demonstrated that key propositions are even possible, it can never directly observe most of its subject (the vast story of life evolving from simple microbes into everything else), and what few practical benefits might be attributed to it are consistent with creationism, too. So in this and similar areas of science, any propositions or positions which are outside the currently-accepted orthodoxy must be regarded as dangerous. Anything which lessens the unanimity of consensus threatens its main strength.

Of course, the resistance to unorthodox geological and evolutionary ideas seldom came close to that which continues to be raised against creationists. Academic degrees have been withheld, jobs lost or denied, careers ruined. All of that deserves to be told more fully, and has been told in the documentary movie "Expelled," and a series of books starting with *Slaughter of the Dissidents* by Dr. Jerry Bergman.

But wait a minute! If Darwinism was "triumphant" and so deeply entrenched that it resists variations, why has the battle with the creationists gotten hotter than ever? The evolutionists would have you believe it is merely because creationists continue to cling stubbornly to their ignorance and religious notions. To those of us on the outside of the establishment, it's because Darwinism was based on a philosophy that has become outdated, the purported evidence for it continues to be far less than convincing, much of it has turned out to be flawed or faked, and the most important steps have still not been demonstrated to be possible.

Section V: Failures of Evolutionism

As creationists see it, the triumph of evolutionism was simply a matter of winning a popularity contest. Think of the story of The Emperor's New Clothes. Now imagine that the con men, instead of going directly to the emperor, got themselves hired as deans of the leading fashion schools. Then, rather than a conspiracy among people planning a hoax, there was an underlying philosophy that the leaders strongly believed. All it takes is for a few experts or leaders of society to proclaim that something is true, and that anybody who doesn't believe it is stupid, ignorant, old-fashioned, or otherwise socially unacceptable. If they base this claim on a popular belief (such as the immutability of natural laws) it acts as a social tuning fork, a shared concept which compels unity. If enough people take up the cry, the movement snowballs and eventually any contrary belief does become unacceptable to the majority of the society.

Failure #1: Failure to convince leading skeptics

The power of this effect may be seen in the fact that Darwinism was opposed by notable scientists as well as conservative religious leaders. None of these scientists were ever convinced by the arguments and data set forth by the evolutionists. They were simply drowned out by the voices of those jumping on the evolutionary bandwagon. The evolutionists controlling our schools and mass media will seldom mention that there was opposition to Darwinism from leading scientists.

(The following list is partly taken from *Men of Science – Men of God* by Henry M. Morris, copyright 1988, 13th printing, 1997, although many sources will provide the same information.)

Richard Owen was not a Biblical creationist, but he was a proponent of a version of the Fixity of Species concept and an anti-Darwinist. What is most significant about his opposition is that he held perhaps the most prominent and influential position in British science at the time, but his opposition was still ineffective. The early evolutionists didn't let authority or consensus stop them, but evolutionists today expect the power of authority and consensus to stop those who question them.

Michael Faraday (1791-1867) was one of the leading scientists of Darwin's time, and one of the greatest ever. He made great contributions to the study of electricity and other areas of physics. Although many consider any sort of belief in creation to be totally contrary to all science, Faraday was a member of a religious group that today would be considered fundamentalist. He continued to believe in the literal, six day creation described in Genesis as he studied the natural world until he died, not at all persuaded by Darwin's *Origin*.

Louis Pasteur was also one of the greatest scientists. While Darwin was dreaming up stories about what might have happened in the past, Pasteur was doing scientific research and experiments which led to great practical benefits. Along with other achievements, Pasteur made great contributions to organic chemistry, developed the vaccine for rabies, and performed a series of experiments which finally proved beyond doubt that not even the tiniest, simplest living thing can form from raw materials alone. After this, evolutionists could only cling to their faith that some unknown set of conditions and combination of materials very different from what exists on Earth today might still somehow produce life.

John William Dawson, LL.D., F.R.S., C.M.G., "the leading creationist in North America during the 19th century," was a Canadian geologist who had done some research with Lyell. Having received high honors and being "the only man to serve as president of both the American Association for the Advancement of Science and of the British Association for the Advancement of Science," he was also "the most distinguished anti-Darwinian in the English speaking world." He outlived Darwin and still held "to the biblical supernatural assumptions." (Page 97 of *The Faces of Origins: A Historical Survey of the Underlying Assumptions from the Early Church to Postmodernism* by David Herbert, M.A., M. Div., Ed. D.; D & I Herbert Publishing, London, Ontario, 2004.)

Perhaps the most powerful example was pointed out to me by Karl Priest: that of Jean-Henri Fabre. He was a true scientist, rejecting the imaginative speculations of evolutionism and sticking to observations and experiments. He studied and did work in several areas, but his main area of work was entomology, the study of insects. He made special investigations into their instincts. He corresponded with Darwin, who had a high regard for Fabre's work at the time, and cited Fabre's work in his book. Fabre, however, saw his direct observations as clearly inconsistent with Darwin's speculative theory that somehow managed to explain everything without making any firm, reliable predictions. Fabre outlived Darwin, and while he respected Darwin personally and for his studies of nature, Fabre remained opposed to "transformism" (as evolutionism was also known). After Darwin died, Fabre noted that Darwin missed seeing that his prediction for the results of Fabre's continued studies had failed. ("Darwin, a true judge, made no mistake about it. He greatly dreaded the problem of the instincts. My first results in particular left him very anxious. If he had known the tactics of the Hairy *Ammophila*, the Mantis-hunting *Tachytes*, the Bee-eating *Philanthus*, the *Calicurgi* and other marauders, his anxiety, I believe, would have ended in a frank admission that he was unable to squeeze instinct into the mould of his formula." http://www.insectman.us/articles/karls/fun-with-fabre.htm, citing https://en.e-fabre.com/virtual_library/more_hunting_wasp/chap11.htm)

Failure #2: Failure to produce useful fruit

One failure of evolutionism which is often overlooked, even by creationists, is that it has not produced any practical benefits. This was one of the benchmarks that Francis Bacon had proposed for judging the claims of science – that eventually they should lead to practical, reliable benefits.

Almost all other areas of science have produced remarkable improvements in technology, transportation, medicine, agriculture, industry, or some other area with practical applications. Astronomy might seem to be an exception, as the celestial objects astronomy studies are very far from us, while other sciences study the world close to us. Yet astronomy has contributed to navigation, timekeeping, and space exploration. The areas of geology concerned with the age of the Earth and evolution (paleontology, etc.) have also been of no practical use. They are sometimes referenced in exploration for metals, oil, gas, and other valuable deposits, but it is the other aspects of geology, which don't require evolutionary and old-Earth assumptions, that are required for such purposes.

While evolutionism is largely concerned with the distant past, its subject matter is all living things, present as well as past. Consider that it is part of biology, which had begun to be studied long before Darwin, and is connected to medicine, agriculture, and other important aspects of society. Consider that it is also claimed to be a fundamental and vital part of biology, or even of all science. Considering all this, is it unreasonable to expect such a thing to rapidly produce major benefits?

Yet over a hundred years after Darwin's theory took the science community by storm, there aren't any practical benefits that evolutionists can point to and honestly claim could not have come about without their theory. In fact, the one benefit most commonly claimed – the recognition of organisms becoming resistant to poisons such as antibiotics, insecticides, and herbicides – is not really a benefit of evolutionism, because evolutionists didn't use their theory to predict it and thereby prevent it or otherwise deal with it beforehand, nor is the phenomenon unexpected within the creation framework. We know in a number of cases the evolution involved is only in the most technical sense – a shift in the ratio of existing genetic variations. Creationists can see and accept the effect as well as the evolutionists. Also, creationists point out that the changes in populations or mutations in organisms involved do not involve any increase in the complexity of any organism, and so the case fits within the creation framework without any problems.

I would like to expand on this point with some examples. I've already noted Pasteur's studies and their benefits. Also during that century, William Thompson (Lord Kelvin) was working on the laws of thermodynamics that are truly fundamental and essential to many aspects of physics and engineering. George Boole began his work on mathematical logic which later proved so useful for computer science. Rudolf Virchow worked on the importance of cells in relation to health and diseases. Joseph Lister developed and promoted antiseptic surgery practices. Gregor Mendel performed his studies of heredity which presented a far different picture from Darwin's mistaken idea, and apparently were so far ahead of their time that nobody understood their value. The principles have since become important aspects of science and agriculture. Mendeleyev formulated his periodic law for the classification of the elements which greatly advanced the understanding of chemistry.

We might also consider the beneficial work of non-scientific men who also lived in the 19th century. All of these were primarily concerned with spreading the Gospel, but they also contributed to social causes. George Mueller was a German who moved to England and started an orphanage the same year Darwin returned from his voyage on the Beagle. Operating on faith alone, without the usual fund-raising techniques, Mueller had opened several more orphanages by the time Darwin published his *Origin*, providing education and care for hundreds of orphans. David Livingston combined mission work with exploration of Africa that was honored by the Royal Geographical Society. The Salvation Army has its roots in the 1860s, as the ministry of William Booth and his wife Catherine. While other people were arguing that we should let natural selection take its course and "decrease the surplus population," or help it along by restricting the "breeding of the lower class" and other outcasts of society, the Salvation Army was providing these unfortunate people with care and comfort.

Getting back to the scientific aspects, consider the fruits of areas of science which didn't appear until later. Edison's phonograph of the 1870s eventually developed into high fidelity record players, and new technology gives us CDs and digital music players. His improved light bulbs have been greatly improved and lighting technology includes advanced fluorescent lights and solid-state LEDs. Tesla's alternating current provides the power for our electronic marvels. Early combustion engines and "horseless carriages" have been improved to the point that vehicles can travel hundreds of miles per hour. Controlled, powered, heavier-than-air flight wasn't achieved until the 20th century (1903), and now there are aircraft that can travel several times the speed of sound. Successful manned jet- and rocket-powered aircraft weren't achieved until World War II, and just a few decades later men had walked on the Moon. Likewise the huge, semi-mechanical computers of WWII which finally fulfilled the plans Charles Babbage envisioned in Darwin's time were improved within a few decades to the range of laptops and supercomputers we have today. Many more examples could be given in all areas of science and technology – the telephone, movies, television, etc.

In comparison, evolutionary studies seem to be stuck in the 19th century. Research has not shown natural selection to be as powerful and strict as needed for the theory, nor have mutations been observed to provide the working material needed. Actual observations have shown that together they may produce a sort of speciation which is consistent with the creationist's framework. That is, the speciation is generally a separation of varieties within a larger, more variable population. This is all that evolutionary studies have found in living things after decades of studying many organisms that go through many generations in the time they are studied, sometimes while being exposed to factors, such as radiation, which speed up mutation rates. Evolutionists remain quite content to have faith in a process that isn't expected to produce one major step under scientific observation.

Failure #3: Fostering Flaws, Fakes, and False starts

This is something which perhaps does not directly bear on the question of the validity of evolutionary theory, but it is certainly closely related and powerfully suggestive. This is the evil fruit of evolutionism. If the worth of a science can be judged positively by its eventually producing practical benefits, may it not be judged negatively if it is especially prone to error and associated with social evils? While all areas of science have seen false leads and charlatans, as well as harmful applications, it seems that evolution, more than any other area of thought claiming to be scientific, has been supported by weak or flawed research, errors, frauds and hoaxes. It was used in the promotion of needless surgeries and has been the basis for bloody social movements. All this may seem too negative to be true, but I will now provide a list which you can research for yourself.

One of the earliest major boosts to evolutionism was the research by Ernst Haeckel that appeared to show that animals passed through the stages of their evolutionary history as they developed from single fertilized egg cells to hatching or birth. Although Haeckel saw evolution as more progressive than Darwin, Darwin was happy to cite Haeckel's work as strong support for his theory. It must have been very hard to doubt evolution when looking at Haeckel's illustrations showing various embryos passing through almost-identical stages. Haeckel's theory of development was known as Embryological Recapitulation, or the "biogenetic law," and described succinctly by the phrase "Ontogeny recapitulates phylogeny." They all sound very impressive, like scientific fact, don't they?

It turned out that Haeckel was wrong, and his drawings were misleading, arguably fraudulent. You may still find similar illustrations, and references to human embryos having gill slits and tails, but embryologists have known from the first that such drawings and statements are oversimplifications at best. Haeckel himself was forced to admit that his drawings were not true to life, but defended them with the poor excuse that he was simply trying to make his point clear to laymen. What look somewhat like gill slits in human embryos are just folds in the skin that form because different parts of the neck develop at different rates. The appearance of a tail is merely the result of the backbone forming faster than the legs and hips develop. Even some of the similarity that does exist is produced by different patterns of development. That such misleading appearances have been used to convince young children of the truth of Evolutionism is scandalous, but few scientists and educators are willing to acknowledge it.

Even the valid data said to support evolutionism has been less than impressive, unless you are impressed by vague authoritarian statements about the sheer volume there is of it. Darwin's finches* themselves were enough to counter the strictest view of Fixity of Species, but as they are all finches, they could easily all belong to one larger group, originally created with that much capacity for adaptive variation. As further study has shown, various species of Galapagos finches can interbreed and produce viable, fertile hybrid offspring. (*Technically, as it turns out, they are no longer classified as finches, but the name has stuck.)

A lot of people have been impressed by charts showing the horse series of fossils, presented as a neat progression of clear evolutionary steps. It eventually came to light, though, that the fossils were not found in such a neat order and there is no solid ground for assuming that there was such an evolutionary progression. Then, too, most of the assumed changes would have been relative, and relatively inconsequential – mainly changes in size, proportions, loss of ribs and toes. Such changes fit within the possibilities of a single created population of animals and don't include the sort of increase in functional complexity that needed to support faith in Evolutionism.

Perhaps the most famous and celebrated case of support for Evolutionism was the peppered moth. The original study claimed to show that birds preying on moths that rested on tree trunks had provided the natural selection required to change the population from one color variation to another. (If mostly dark and mostly light can be called color variations.) There were some serious flaws in that study, which eventually caused quite a scandal, but that is irrelevant. The important thing is that the case (and a modern repetition that was done correctly) was used to convince people that evolution of one kind of animal into another had been observed to be possible, but it has not. The elimination of a color variety from a population, or even so much as the appearance of a new color type due to a mutation, does not demonstrate the sort of change that is required to go from scientific evidence for evolution to support for Evolutionism. Anyone who makes the effort to think clearly and rigorously can see that there is no increase in complexity in a color change, and that no matter how many such changes accumulate, they won't add up to the kind of change that would have to occur to account for all living things evolving from microbes.

The history of evolutionary theorizing is filled with proposals that enjoyed a brief popularity and later were dropped like hot potatoes. One of my favorite examples is the idea that insects developed flight by beginning with fixed-wing gliding (see *Discover*, "The Physics of. . . Insect Flight," by Robert Kunzig, Saturday, April 01, 2000, p. 27).

Some of the worst examples put forth in support of evolutionism have been seen in the efforts to include humans in the picture. In the early years, there weren't any fossil apes to propose as possible transitional forms, but Neanderthals and Cro-Magnon man were forced into that role. These "cave-men" were depicted as brutish or even ape-like, and at best as primitive, dull-witted savages. These images still color the public imagination, although much evidence has since been found showing that Neanderthals were capable of producing artifacts which modern humans find challenging to replicate.

There have also been a few major embarrassments in this area. One proposed "missing link" was popularly known as Nebraska Man, made famous by its discovery about the time of the Scopes trial, and by an illustration in a newspaper with wide circulation. It was presented by a distinguished professor as proof positive that humans had evolved from apes, and thus a reason for the opponents of evolutionism to be ashamed and give up the fight. That was a heavy burden to place on the fossil, which consisted of a single tooth. It wasn't long before further research revealed that the tooth had belonged, not to a human, ape, or ape-man, but to an extinct wild pig.

Of course, amazing discoveries get better press than corrections and retractions, so the imaginary Nebraska Man had done his part to support evolutionism. Besides, the evolutionists could point to another shining example of a fossil having some human and some ape-like traits, and it consisted of a complete skull and more. This discovery was popularly known as Piltdown Man. Unfortunately for the evolutionists, it was later discovered to be a much greater embarrassment than Nebraska Man. It wasn't merely a case of misidentification, it was an outright hoax! Somebody had modified an ape jaw and planted it at an archaeological dig along with a human skull and other material. Most unfortunately, while it had been discovered in 1912, it wasn't definitively proven to be a hoax until 1953.

There were other fossils of extinct apes and apparently primitive humans, but none of them at the time seemed to qualify as being very far into the gap between. Then in 1959 came the discovery of "Zinjanthropous," nicknamed "Nutcracker Man." This was used through the 1960s and into the '70s as an example of an ape-like creature that could shape stones into tools and could very well be our ancestor. Although this was presented as an unquestionable fact to the public, as early as 1965 evidence indicated it was not a toolmaker and was clearly not one of our ancestors. (See: Reader, John, "Whatever Happened to Zinjanthropus?" *New Scientist*, 89:802, March 26, 1981)

Finally, a genus of ape (*Australopithecus*) which had been discovered in the '20s was dusted off with the discovery in the '70s of a new fossil dubbed "Lucy," and, based on indications it could walk upright better than other apes, was given the most human-like possible depictions and has served as the main "missing link" ever since.

And so, by presenting first one fossil and then another as "the missing link," emphasizing the more human-like aspects of fossil apes and the supposedly primitive features of fossil humans, evolutionists conjured up the impression of an entire evolutionary lineage from apes to humans. Technically, with one or two exceptions, it was supposed to show a series of "sister species" of the common ancestors shared by modern apes and humans, but generally such technicalities weren't shared with the public at large.

An illustration of "The Ascent of Man" became an icon that is still familiar enough to spawn parodies, but even some evolutionists admit it was misleading at best. Is it any wonder that creationists keep referring back to it? According to evolutionists, we should let such past things be forgotten, and anyway, all this merely shows that science corrects its own errors. The point the creationists won't let go of is this: If a certain area of science is constantly correcting itself, discovering that evidence set forth as reason to believe in it had been false, and sometimes having to correct its corrections, why should we put any confidence in the claim that it is correct now? Furthermore, we see a pattern of large initial claims being discredited by further research. How many miscues does it take before the situation becomes reminiscent of Bullwinkle Moose trying to pull a rabbit out of a hat, saying, "This time for sure!"?

Although Darwin never publicly tied his theory to the origin of life, consistent application of the mindset that leads inevitably to evolutionism also leads to the conclusion that somehow life evolved by natural forces acting on raw materials. Yet this is arguably the greatest failure of evolutionism. Without proof that life could evolve in the first place, biological evolution can't get to square one, and so arguably doesn't deserve further consideration. If one allows that life may have begun supernaturally, then the playing field (including public education) should be opened up for creationism, and evolutionists refuse to allow that.

It isn't as if evolutionists haven't tried to show life evolving from inanimate chemicals. Ernst Haeckel and Thomas Huxley postulated that some kind of proto-life probably still existed. Haeckel called it "Monera" and drew illustrations of what it would have looked like. When a research ship dredged up some odd sludge, the belief that it was some form of proto-life lasted long enough that it was given the name "*Bathybius haeckelii.*" However, it was just a chemical sludge that wasn't anywhere near the hypothetical path to becoming alive.

I indicated that evolutionists have tried to show that life could arise by purely natural forces, but for a long time, they didn't even try, at least not enough to draw notice. Still, without any evidence that it could happen, students were taught that life arose from what was called "the primordial soup" of chemicals that was supposed to have more or less covered the early Earth. The original concept of the primordial soup has been consigned to the large pile of evolutionary conjectures and supposed evidence that didn't pan out. Variations continue to be taught, but ultimately it is a statement of faith, not science.

Some evolutionists finally stopped conjecturing and started experimenting, trying to discover exactly what it would take to turn a natural combination of chemicals into something that could honestly be called a living thing. The first attempt (if there were earlier ones, you're not likely to hear about them) was in 1953, when Stanley Miller performed an experiment as part of his PhD thesis, under the direction of Nobel Prize winner Harold C. Urey, which he hoped would simulate the conditions on the primordial Earth. When the experiment produced amino acids, a class of chemicals which are vital parts of living things, it wasn't long before headlines and news stories celebrated the tiny success as a major advance, and predicted that further great strides would soon be made. There were optimistic articles claiming that it wouldn't be long before scientists created life in a test tube. The Miller-Urey experiment continues to appear as evidence for the evolutionary origin of life. The full list of its many shortcomings is not as likely to be included.

Besides the obvious fact that no living thing has ever been produced by such experiments, the chemical composition Miller used is no longer believed to accurately reflect the mix found on the early Earth. It is also common to leave out the fact that the equipment included a section which trapped chemical products of reactions so that they were protected from further exposure to the conditions that had created them. Another point which may be left out is that the amino acids produced in any one experiment include only a few of the ones found in living things, and some of the other products would be deadly if any life happened to form anyway.

Furthermore, in living things all the amino acids have a certain arrangement or handedness (the technical term is "chirality," and amino acids in living things are all left-handed), whereas Miller's experiment produced equal amounts of both, as is found when amino acids are made normally in a lab; and when living things die, their amino acids decay and assume equal handedness. A similar chirality challenge exists for the sugars in living things, but their arrangement has the opposite chirality. Oh yes, that's another point – Miller's experiment was aimed only at amino acids, but living things require other kinds of materials, such as sugars and lipids.

Miller continued to work on such experiments, and others have tried quite different experimental conditions, but no one in the decades since has been able to produce life. In fact, no one has gotten much farther than that first experiment. Various experiments have produced all, or nearly all, of the amino acids found in living things, but each experiment only produces one or a few of them. No experiment has both simulated natural conditions and produced a large number of the amino acids needed for life. No experiment has demonstrated how natural conditions could produce a chemical or set of chemicals that could duplicate a string of molecules carrying a code like that found in DNA. No experiment has shown that natural conditions could produce a set of amino acids that all have the same chirality. Evolutionists simply continue to have faith that this is just one more gap in our scientific knowledge that will eventually be filled. Is it any wonder that creationists consider evolutionism a kind of religious belief? Is faith in an evolution-of-the-gaps any better than faith in a "God of the gaps"?

Another case of an evolutionism-related misstep in science appears to be that of Junk DNA. An early evolutionary belief was that humans, being at the end of a long line of evolutionary progress, had a lot of parts (vestigial organs) that were useless leftovers from our animal ancestors. This argument is still sometimes advanced as support for evolution, but the list of apparently useless organs has shrunk from almost 200 to about 2 (depending on who's counting), and the ones left are debatable. Still, evolutionists believed that much of our DNA was vestigial, useless junk left over from ancestral genomes. Creationists believe that the human genome is degenerating at an ever-increasing rate, but this would not have already produced as much junk as evolutionists were proposing. However, the proposal may easily have appeared to be data-driven, as scientists began researching DNA and finding that much of it did not provide codes for any proteins. Yet they might have considered the DNA had other purposes. It is now becoming clear that the expectation that there would be a large amount of useless DNA was premature. While evolutionists were beginning to grudgingly note some cases of "junk" DNA being used for purposes other than protein codes, creationists predicted that we could expect many more cases. Sure enough, many examples of what were thought to be junk sequences have turned out to have important functions, especially in controlling the activity of the protein-coding genes.

Evolutionism has also had some negative practical applications. The idea of vestigial organs resulted in huge numbers of surgeries that weren't needed. It inspired the idea that our tonsils and appendix were worse than useless, that they were dangerous remnants of the past, and should be removed routinely, even when nothing was wrong with them. Further research has shown that the tonsils contribute to the body's immune system, and the appendix helps maintain the proper balance of beneficial micro-organisms in the intestines.

Another case of evolutionism influencing perception of human health reveals a darker side. Based on Haeckel's notion of embryologic recapitulation of evolutionary history, Down's syndrome was originally named (by Dr. Down) "Mongolism." Dr. Down believed these individuals features indicated they were arrested in their evolutionary development before they gained fully Caucasian status. (See "Dr. Down's Syndrome," by Stephen J. Gould in Natural History 89, April 1980).

Arguably, it is unfair to saddle a science with all the unscientific misdeeds perpetuated by those inspired or encouraged by it. On the other hand, what other area of science has had such an effect? Darwinism has had little real influence on society through the usual channels of science, but a very large and very negative influence by way of philosophy and politics. This provides another reason to consider evolutionism a sort of philosophy or religion rather than science. As mentioned earlier, cutthroat capitalism, European colonialism, and social Darwinism immediately seized on Darwin's view of nature as justification for their excesses. Marxism and Communism also embraced Darwinism as justification for the revolutionary overthrow of the old order by the new. (Those who are fussy about such matters may wish to recall that the Soviet Union substituted their own view of evolution, a Lamarckian scheme called Lysenkoism.)

The darkest deeds in history were committed in the name of eugenics, a Darwinism-inspired and supposedly scientific practice. Francis Galton, a cousin of Charles Darwin, coined the term and promoted the concept with the full knowledge of Darwin, although there seems to be little material available showing what Darwin thought of it. At any rate, there is also no evidence that he strongly objected to it or thought it was a misapplication of his theory.

After all, eugenics is based on Darwin's conception of natural selection and his analogy with artificial selection. According to eugenics, Darwinism teaches us that even the slightest negative traits should be selected out of a population, and that great advances can be made by diligently breeding only the best members. This intense selection of minor variations was exactly what Darwin claimed had driven life from humble beginnings all the way up to humans. His *Origin* has many statements such as these: "...each new variety...will generally press hardest on its nearest kindred, and tend to exterminate them." "...natural selection will separate all the superior individuals...will destroy all the inferior individuals." "... no one can solve the...problem, why, of two races of savages, one has risen higher in the scale of civilization than the other; and this apparently implies increased brain power." The implication for society is clear: allow the less fit members of society to reproduce freely and the human race is in danger of degeneration; apply strict breeding programs and the race will evolve into a superior species.

Is it a very large step from there to Hitler's vision of Aryan supermen? The Nazis are commonly portrayed as right-wing throwbacks to the Dark Ages, but that is quite contrary to the historical facts. The Nazis were, after all, the National Socialist party, not the Conservative Capitalist party. They saw themselves as leading the way in the world's progress, and strongly promoted science. Indeed, Nazi Germany led the way in aeronautics in the 1930s, and in jet aircraft and rockets by the end of the war. There were also many scientifically-minded, progress-seeking men outside the Nazi party and Germany who agreed with the basic philosophy, although most did not advocate Hitler's drastic concept of forcefully eliminating Jews and other people they believed to be "inferior stock."

Of course, social injustices and racism existed long before Darwin. The point here is only that Darwinism not only provided a new justification for such things, it bestowed on them the appearance of being inevitable, good, progressive and scientific – the means to a bright future for the superhuman race our descendants would evolve into.

This may be an especially hard thing for many people to accept, but review the quotes from Darwin's *Origin* again. Consider examples of the "scientific" attitudes of Darwin's day that included viewing dark-skinned modern humans in other cultures as primitive, animal-like – living cave-men or not-so-missing links. Examples include Darwin's own negative assessment of certain South Americans (which, to be fair, he somewhat retracted after seeing the results of missionary efforts) and references to "savages" in his writings. The hunting of Tasmanians and Australians, and displaying skeletons and corpses of "primitive" people in museums were also justified as advancing scientific knowledge. Remember, Darwin attacked the notion of Fixity of Species by pointing to new varieties of livestock being as different from the old as different species in the wild are from each other. So it was quite natural for evolutionists of the time to believe that humanity should be divided up into several distinct races of unequal evolutionary development, some of which might not be considered fully human.

The existence of this supposedly scientific racism should be beyond doubt, but it doesn't get much press, and little, if any, coverage in textbooks. In the review article "Painting the whole picture?" referenced earlier, the reviewer notes Haeckel's racism and its contributions to the rise of Nazi ideology, noting that "...one could argue that Haeckel's dark side was as much a natural consequence of his world view as was *Art Forms in Nature.*"

Doctor Larry Azar notes in his book *Evolution and Other Fairy Tales* (p. 205) that the greatest promoter of Darwinism in England was also a racist: "Moreover, for the Negro specifically, Huxley asserted that he had 'not the smallest sympathy' for him, and cautioned, 'Don't believe in him at all.'[81]" Perhaps even worse, "Huxley...insisted that 'No rational man, cognizant of the facts, believes that the average Negro is the equal, still less the superior, of the white man.'[82]" (81. Huxley: *Life and Letters*, I, p. 272.) (82. Huxley: *Lay Sermons*, p. 20.)

H. G. Wells studied under Huxley. "Wells advocated a level of eugenics that was more extreme than Hitler's. The weak should be killed by the strong, having 'no pity and less benevolence'. The diseased, deformed and insane, together with 'those swarms of blacks, and brown, and dirty-white, and yellow people ... will have to go' in order to create a scientific utopia." ("H.G. Wells: Darwin's disciple and eugenicist extraordinaire" by Jerry Bergman, *TJ* 18(3): 116-120, December 2004)

Americans should keep in mind that the doctrine of eugenics was also popular here, and was applied to some extent. The forced sterilization of some people was approved by the Supreme Court, with judge Oliver Wendell Holmes presiding, in 1924 (as noted in Creation 19(4), Sep-Nov 97, p. 22). This was the same year that Hitler wrote *Mein Kampf* and one year **before** the Scopes trial. The biology text that Scopes allegedly used to teach evolution also promoted eugenics.

While the horrors of Hitler's genocidal policy turned most people away from eugenics, the underlying evolutionary principles are still taught as fact. Sometimes the old attitudes still peek out from under the scientific covers. Probably the exciting prospect of studying living examples of "primitive cave-men" was a major factor in the success of the "Tasaday hoax." The Tasaday were a tribe of people discovered in a jungle in the Philippines in 1972, still living a life so simple and primitive as to hardly seem human. They were featured in *National Geographic* and in a television special. Later expeditions, however, found no trace of the tribe – except for some individuals living just like everyone else in the area, who confessed to participating in the hoax. Now it appears they were indeed a separate tribe earlier, but it is clear that they had not been isolated for as long and as completely as the early reports claimed, nor did they live in such an extremely primitive manner.

Some evolutionists are repulsed by certain aspects of modern applied evolutionism. Many reject the approach and conclusions of sociobiology, in which beliefs and guesses about the evolutionary path of our ancestors are used to explain our own behaviors. For example, some sociobiologists proposed that men are prone to raping women due to the evolutionary success of ancestral species that had multiple mates.

Failure #4: Fixed to a Flawed Foundation

The most fundamental failure of Evolutionism would have been hard to see until other areas of science advanced. While other areas of science went on to produce bigger and better things, evolutionary theory is still splashing around in the shallow end of the pool and making messes. But it's worse than that. Evolutionism is based on the premise that nothing exists except the shallow end of the pool, and other areas of science have shown that there is a deep end, connected to a whole ocean beyond that, which the scientists of Darwin's day didn't know existed, and even denied that it could exist.

To understand that strange metaphor, think back to the beginning of the chapter, especially Subsection C: The Universe as a box of billiard balls. Evolutionism could only thrive and dethrone belief in divine creation and a global Flood when the scientific data seemed to fit (or was made to fit) the philosophy of absolute naturalism, or at most a form of deism that denied any miracles or other supernatural interference with the laws of nature. Practically speaking, there's no difference, as either way, the entire universe had to be viewed as a closed box of ordinary matter and waves of energy, from start to finish and from the largest to the smallest things in it. Everything in it must have always obeyed the same laws of physics as we see and measure them today, without exception, and without any great departure from known rates of activity or change.

This philosophical box, constraining what scientists were allowed to consider (at least while practicing their art) is what I call the shallow end of the pool. It doesn't deal with the hard science of the scientific method. It is the faith that all of Reality is a box of ordinary things behaving in ordinary ways, or if there is anything that might be thought of as above, beyond, beside, or otherwise outside that box, it would never alter the ordinary physical course of things inside it. It was required for the triumph of evolutionism, because if science allowed for anything outside the ordinary, then all bets were off as far as scientists convincing others that their work was like a time machine, revealing all sorts of unquestionable facts about the unobserved past. It was also the basis for the new, atheistic rationalization for the foundation and reliability of science. The uniformitarian geology that cradled Darwin's vast ages of slow, gradual modifications especially depended on having faith that there had never been any extraordinarily great changes in the rates of natural processes or major unknowns in the physical forces at play in nature. Either one would have weakened the apparently solid case laid out by Hutton and Lyell, even within the naturalistic philosophy that had become part of the new definition of science.

In the decades before and immediately after Darwin's theory was published, most of the advances of science seemed to confirm the view that the universe was a closed box, a shallow pool. Pasteur's demonstrations destroying the idea that living things would naturally form spontaneously in a nutrient broth cast a wet blanket over the idea that the closed box of billiard balls could ever have produced life. However, the evolutionists were content not to press the most controversial points (even Lyell was repulsed by Darwin's extending his theory to include humans) until after they'd established their dominance over the few relevant areas of science.

But by the end of the 1800s, there were already hints that the box wasn't so solidly closed, that there might be much more to the world around us than the little pool. As the 20th century dawned and progressed, evolution's foundational picture of the universe was washed out by other areas of science, leaving the edifice to stand by virtue of its monopoly over academia and scientific organizations; by authoritarianism, the high prestige of science in general, and a tissue of data points that included irrelevant, flawed, and false evidence. It was a real, modern case of The

Emperor's New Clothes, shingles of faith in "facts" supported by a hidden fog of false philosophy.

Very few people, if any, recognized this undermining of evolutionism for what it was until late in the 20th century. This is not surprising, because the changes came slowly, in scattered areas of science, and the connection or significance could only be seen by someone familiar with the underlying philosophy. It's easier to see when reviewing the history in hindsight.

The first hints came in the closing years of the 1800s, with the discovery that atoms weren't solid and eternal, but could break down into smaller parts and radiate energy. Max Planck formulated quantum theory in 1900, and Albert Einstein soon put forward his Special Theory of Relativity and a paper treating light as packets of energy and not continuous waves. Later work suggested that matter could likewise be treated as having aspects of waves, that matter could be converted entirely into energy and energy could form into particles of matter. Still later research supported aspects of quantum mechanics that even Einstein found too strange to be comfortable with. Electrons can cross barriers which would be impassable according to classic physics. Examining a particle that has a "quantum entanglement" with another particle instantly determines the characteristic examined for both particles, no matter how far apart.

The relevance here is that the Bible's record of the past had been rejected in favor of the interpretations of natural phenomena by scientists, not because of scientific evidence on its own merits, but from assumptions promoted by philosophers who pointed to how ordinary, regular, and slow in changing everything in the world seemed to be. The materialistic box seemed so absolute and the world ultimately simple. In contrast, how could God be One and Three at the same time? How could Jesus violate the laws of God's creation to work miracles? How could he enter a locked room? The closed box of bouncing atoms view of reality did not allow for such things. When this philosophy infected science, it seemed to work, opening up new areas of science, through extrapolating the ordinary indefinitely. But as time went on, the original way of doing science (sticking to observations and experimentation) began to reveal that our "ordinary" world is stranger than those philosophers ever dreamed.

The developments in subatomic physics were paralleled in cosmology. The picture of the universe began changing when it was determined that some objects that had been taken as clouds of gas and dust within the Milky Way (considered then to be the entire universe) were far distant "island universes" – what we now call galaxies. Advanced astrophysics led to the prediction of the existence of Black Holes, places where the collapse of massive stars (or collisions of stars) created such intense gravity that space and time were warped, trapping even light. Beyond the "event horizon," the very laws of physics as we know them would be distorted beyond recognition.

Observations indicating that the entire universe was expanding in all directions led inevitably (according to naturalistic philosophy itself) to the conclusion that at some point in the past, the entire universe had been just a tiny ball or point of unimaginable energy. At least one theory suggests that many times more energy existed at first, but as it cooled and formed nearly equal amounts of matter and anti-matter, the two canceled each other out and somehow only a relatively small amount of matter escaped the annihilation – and that was all the matter that now makes up the universe.

Other research suggested that there must be vast amounts of matter in the universe that we can't see, apparently having no detectable characteristics other than its mass. Understandably, it is called Dark Matter. More recently, study of the universal expansion rate has suggested that some force, dubbed Dark Energy, is acting as anti-gravity, increasing in strength as the universe expands and causing it to expand even faster. As of this writing, there is still no direct observations of these exotic forms of matter and energy, or knowledge of what they are, besides the effects which they are assumed to be causing. Yet, if they do exist, they must make up something like 90% of the total mass and energy of the universe.

These developments show that many things once thought to be scientific certainties are not certain after all. They show that the universe is bigger and stranger than we knew. They point to things beyond the physics of our observed universe. Our once seemingly simple little box of a universe began, even according to evolutionists, in a way that could be considered magical or miraculous, it is sprinkled with holes, and apparently filled with matter and energy unlike anything we are aware of in our ordinary environment. In a universe like this, it should be obvious that ruling out the possibility of supernatural events is just a philosophical preference.

Even in more down-to-earth matters, science has been moving farther and farther from uniformitarian doctrine. In 1912 Alfred Lothar Wegener developed a theory of continental drift based on fossil and glacial evidence. This drift was a very slow movement, but his theory was ridiculed until long after his death because he could provide no adequate explanation for the cause of the phenomenon. Furthermore, even this slow but very great change in geology didn't fit the extreme uniformitarianism of the time. Plate tectonics, with continents colliding to form super-continents, and breaking up again several times throughout Earth's history, is now the standard model of geology.

Then there was the "Great Scablands Debate." In 1923 J. Harlen Bretz proposed that a huge flood explained how a vast area of unusual geologic features had formed in Washington state. He had to suffer the verbal slings and arrows of his colleagues for decades. Proposing such great catastrophes to explain things instead of the traditional accumulation of ordinary effects simply wasn't acceptable. (See: "The Great Scablands Debate," Natural History, Aug/Sept. 1978) It was about forty years before enough other scientists grudgingly allowed themselves to see that the evidence could best be explained by a flood far greater than any recorded by scientists or known in history – outside of the Bible's account and traditions of a global Flood. It would have been merely an after-effect of the global Flood in the Bible, but this and some similar cases seriously undermine the worldview that produced anti-Scriptural geology and evolution.

One other case that shows how the old uniformitarianism has seriously crumbled is the popularity of catastrophic theories about the extinction of the dinosaurs (and other creatures). According to many, it is beyond doubt that a large object from space crashing into the Caribbean was responsible. Other scientists point to large volcanic lava flows in India. There are other theories, and debates over the rate of extinction, but generally there is agreement that some great catastrophe was involved and that the final blow was far greater and happened far faster than the old uniformitarian establishment would have deigned to consider.

It's important to note that in some of these areas of science, the new discoveries, as revolutionary as they were, did not force scientists to toss out everything related and start from scratch. Nothing can nullify repeated observations and experiments properly carried out. Newton's laws and formulas remain valid for physics in our everyday experiences, and Einstein's equations hold up except in the special cases where quantum dynamics are required. Modern physics has opened doors to strange new aspects of reality without losing the ordinary.

It's a different situation for the areas that consist of interpreting data via naturalistic philosophy and nothing more. Evolutionists found themselves constantly tripping over things and having to go back to square one. For decades (starting while Darwin was still alive) the evolutionary bandwagon was magically running on the imaginary fumes of a fuel that never existed. Darwin had proposed a means or process of inheritance that didn't stand up to inspection for long. Nobody else understood how biological inheritance worked. Mendel had begun to discover (by experimentation) the major principle of how it worked, but nobody who read his report recognized its value. It wasn't until decades later that his work began to be replicated. The bad news for evolutionists was that Mendelian inheritance was completely contrary to Darwinian evolution. What Mendel had discovered was that almost all traits and variations were merely combinations of traits already present in the parents or previous generations. The degree of variation was influenced by environmental factors, not inheritable ones.

Evolution was up a creek without a paddle, but evolutionists grabbed the one straw they could find and made that serve. Darwin had been aware that sometimes animals show surprising and strange variations (mutations), but had relegated such changes to a very minor role in his story. When evolutionists discovered there wasn't any other factor in reproduction and inheritance that could provide changes not present in previous generations, they seized on this odd bit player and made it the star of the show.

That fact that Evolutionism is a speculative application of naturalistic philosophy without solid foundation can be seen by observing that even among those dedicated to this philosophy, there have been disputes and variations which the establishment has deemed threatening, and forced them to fit the mold. These incidents are more like the activity of a religion than science.

Some researchers noted that many forms of life show only minor variations throughout the fossil record, in some cases from the lowest strata to living forms. They also noted that the Establishment mostly ignored such cases and played up the few fossils said to show evolutionary transitions. Breaking ranks a bit, these scientists proposed that perhaps evolution was not a constant aspect of life after all. They suggested that, instead of assuming the fossil record didn't accurately reflect the evolutionary process, the theory of evolution should be modified to fit the data.

In this view, called "punctuated equilibria," the most significant steps of evolution must have occurred relatively rapidly, in small and isolated populations, which then afterwards might not evolve any further. The debate over this variation of evolution became very hot, and grew until it spilled out of the halls of academia into the popular science journals. About this time, however, the evolutionists became aware that creationists were pointing out that a lot of the points brought up were very much like what they had been saying all along! Not surprisingly, the evolutionists quickly began to close ranks again, and the two sides adjusted their views until the debate was merely a disagreement over a technicality. Still, the unevenness of the fossil record remains, and we now know there are also discrepancies between the fossil patterns and comparisons of biological molecules, including DNA. The evolutionists have to resort to guessing that there were relatively rapid "adaptive radiations" of amazing breadth in several cases, or long lines of creatures that somehow left no known fossils.

A similar case (although smaller and less well-known) is that of the rise of the classification method known as cladistics. Although evolutionists had taken Linnaeus' classification system (which had been devised to reveal the order within divine creation) and claimed it supported universal common ancestry, some researchers attempting to classify fossils realized that they didn't need to take evolution into account to do their work. All they needed to do was make lists of traits and group the fossils according to the number of traits they had in common.

Here's how Dr. Azar describes the situation (*Evolution and Other Fairy Tales* p. 144): "... cladists... classify organic beings according to genealogy, including phylogenetic branching. This primary occupation of classifying organisms represents a major threat to evolutionary theory. According to one proponent, cladistics is either 'neutral or [even] opposed to evolutionary theorizing.'[65] Another proponent voices the same message: cladistics is 'at odds with evolutionary thinking.'[66]" [notes: 65. Ball: "On Groups," p. 446. 66. Beatty: "Classes and Cladists," p. 25. Also see Oldroyd: Cladistics is "fundamentally a non-evolutionary classification." ("Charles Darwin's Theory of Evolution: A Review of Present Understanding," *Biology and Philosophy*, I, p. 133.)]

Of course, the major journals such as *Science* and *Nature* reacted very negatively to the non-evolutionary implications of cladistics, and its practitioners quickly pledged their allegiance to the dogma of evolution and learned to formulate their endeavor in overt evolutionary terms. Still, you may find hints of the essentially non-evolutionary nature of cladistics, for instance in *Nature* v. 430, 29 July 2004, p. 506, "The body-plan explosion," which also touches on the largest "adaptive radiation" in the fossil record. The patterns attributed to adaptive radiations, and classification analyses producing long branches topped by "bushy trees" (lots of twigs at the ends), take a step away from Darwin's vision of the Tree of Life, and toward the creationists' concept of created "kinds" within which there are many possible minor variations.

Failure #5: Failure to Finish Foes

The plainest contradiction to the supposed triumph of evolutionism is that there have always been scientists who oppose it on scientific grounds. If anything, the scientific opposition to evolutionism began growing stronger in the second half of the 20th century, and it is growing stronger still in the 21st. According to the evolutionists, this is only due to the stubbornness of religious fundamentalists, but anyone who is not religiously attached to evolutionism should be able to see that this is not the case. There are many old-Earth creationists. There are also a lot of people who have decided to do whatever it takes to make their religion compatible with the practically atheistic doctrine of evolutionism. Some people do seem to hold onto beliefs no matter what, but there certainly seems to be much more to creationism than that. If the evidence for Evolution were as convincing as evolutionists claim, there wouldn't be so many creationist scientists; perhaps hardly anyone who studied much science would doubt it, no matter how strong their religious beliefs are.

An excellent documentation of many of the scientists who rejected evolutionism after Darwin is found in *Men of Science – Men of God* by Henry M. Morris. Here I will borrow a few examples from it.

George Washington Carver (1864-1943) is famous for having "developed over 300 products from the peanut and over 118 from the sweet potato." This biochemical wizard "was also a sincere and humble Christian, never hesitating to confess his faith in the God of the Bible and attributing all his success and ability to God." He went on record as having looked to the Creator to help him find the blessings He had placed in His creations.

Douglas Dewar "was a founder of the Evolution Protest Movement in London in 1932... He had been a graduate of Cambridge in Natural Science and was an evolutionist in his early career... He had a distinguished career...as a naturalist and ornithologist... After he became a Christian and a creationist... he wrote numerous papers and books expounding the scientific basis of creationism. ... and participated in a number of ... debates with leading British evolutionists..."

Paul LeMoine "was President of the Geological Society of France, Director of the Natural History Museum in Paris, and a chief editor of the *Encyclopedie Francaise*, 1937 edition." This volume commented on evolution: "The theory of evolution is impossible... a kind of dogma which the priests no longer believe..." even though "LeMoine had once been an evolutionist himself."

George McReady-Price (1870-1963) has been presented as the inventor of "creation science." (See Ronald L. Numbers: *The Creationists*.) As seen in our historical review, it may be more accurate to view him as the last of a line of scientists who were tolerated and even respected within the mainstream scientific establishment in spite of their open religious beliefs. His work (inspired by his Seventh-Day Adventist beliefs) did revive the effort to interpret geology in terms of a relatively recent creation and global flood. (See also Azar's book and "Creation and Genesis: A Historical Survey" by Andrew S. Kulikovsky, *Creation Research Society Quarterly*, Volume 43, March 2007, pp. 206-219)

In the 1950s and '60s, atheistic and anti-theistic sentiments, movements and activities became more public and socially powerful than ever before. Evolutionism was prominent among the battle standards of those proclaiming "God is dead!" and demanding that all religious activities and symbols be removed from public view. Was it just coincidence that at the same time, the education system saw a renewed effort to make the teaching of evolution a major part of science curricula? For decades evolutionists had ignored the few counter-examples and claimed that there weren't any scientists or "scientific men" who doubted evolution. When Dr. Price died, it must have seemed to many of them that the saying had finally become absolute truth.

But there were still scientists who rejected evolutionism, and in 1961, an event occurred which many in the modern Creation Science movement regard as the beginning of the current resurgence. This was the publication of *The Genesis Flood* by Dr. Whitcomb and Dr. Henry Morris. A compilation of creation-supporting observations and arguments (MacReady-Price's work providing much ammunition regarding geology), along with sound exposition of relevant Bible passages, observations of failures of evolutionism, and new suggestions for scientific explanations regarding creation and the Flood, it gave encouragement to creationists and sparked interest and enthusiasm in the scientists who still believed in creation.

In 1963 the Creation Research Society was founded as a professional society of scientists and educated laypersons, from around the world, promoting scientific research and publication supporting special creation. In 1972 Henry Morris founded the Institute for Creation Research, an institution for creationist literature, education, and advocacy. Both of these remain today and hold their original young-Earth position. They have been joined by other creation science organizations in America, Australia, and Europe.

The evolutionists might have ignored the creationists' organizations for much longer than they did, but the creationists challenged the evolutionists like David taking on Goliath. Duane Gish and others challenged evolutionists to debates in universities and other public venues, and the evolutionists (used to thinking of creation as a long-dead myth) were shocked to find themselves losing face badly. Eventually some evolutionists would advise the others not to agree to such debates, ostensibly because the creationists were better debaters and evolution was a subject that did not lend itself well to the debate format. Understandably, creationists see this as a lame excuse to avoid admitting defeat.

At any rate, a significant portion of the public had been educated about the weaknesses of evolutionism and the scientific points of support for creation. This was all the more galling to the evolutionists because at the same time they'd been trying to make evolution the major theme of biology, if not all of science or even education in general. (In 1972 Theodosius Dobzhansky famously claimed that "Nothing in Biology Makes Sense Except in Light of Evolution" at a gathering of the National Association of Biology Teachers.) Suddenly students were asking questions they weren't supposed to, an awkward thing for teachers to avoid in a field that was supposed to be about asking questions. Many of the new creationists were converted evolutionists! Worse yet in the eyes of many evolutionists, scientists who didn't want to be considered creationists were beginning to openly express doubts and questions about evolution.

One of the first of these new dissenters was Michael Denton, whose book, *Darwinism: A Theory in Crisis*, was clearly coming from a new viewpoint. It didn't espouse biblical creationism in any way, but merely suggested that the data did not offer good support for Darwinism, even in its current form. Not very long after, a lawyer named Phillip Johnson looked into the subject and (as described in his *Darwin on Trial*) discovered that the doctrine of evolutionism was being supported more by the tricks of his trade than by real science. In the last years of the 20th century, the Intelligent Design movement began, as 1996 saw the publication of *Darwin's Black Box* by Michael Behe, plus a conference of scientists called *Mere Creation*. Johnson and Behe were joined by William Dembski, Jonathan Wells, and others in forming the Discovery Institute.

Evolutionists were dumbfounded and outraged. No longer able to deny that there were scientists and highly intellectual men who doubted evolution or even believed in creation, their first (and ongoing) response was to disparage everyone who stepped outside their camp, to the point of slandering them and destroying their careers, and claiming that they were all fundamentalist creationists. Yet, while making it clear that they believed that any wavering from the purely naturalistic, intelligence-free evolution dogma must be motivated solely by religious fervor and obstinacy, they also began to point out that evolutionism was perfectly compatible with religion! For example, Dr. Kenneth Miller's *Finding Darwin's God* attempts to reconcile evolutionism and religion – if the religion accepts that mankind evolved from animals, Genesis is a collection of myths, and God didn't do any creative work that couldn't have happened naturally. To a large extent, however, evolutionists have continued to rely on empty boasts and pretending they don't have any scientific opposition. Published in 2000, Niles Eldredge's *The Triumph of Evolution and the Failure of Creationism* pontificated, "the debate has been dead since 1859 – and evolution was triumphant!"

There are lists of creationist scientists and "scientific men" who doubt the established story of Evolution to one degree or another, which you can easily look up on the internet.

Failure #6: Failure to Foster Free Inquiry

Evolutionists know that they won't be able to keep making such claims if students are given any encouragement to question evolutionism and consider that the world might have been actively created. In America, belief in an active, personal God remains strong, but evolutionists can follow the atheists' example of how to do an end-run around the will of the majority. Ever since the Scopes trial, the ACLU and other powerful pro-evolutionism organizations have been using the court system to establish evolutionism-compatible religion as the only kind allowed in public. Their greatest leverage in this effort has been the establishment of a biased interpretation of "the establishment clause" of the First Amendment. They are greatly helped in their crusade by the major media and liberal theologians, both of whom are always happy to promote liberal causes and attack other traditional and conservative religious views. Ironically, while the cry of the evolutionists at the Scopes trial was for freedom of speech and fairness, once they firmly had the upper hand their cries turned to support for censorship and intolerance.

The evolutionists have also been greatly helped in this matter by well-meaning but naive creationists in legislatures and boards of education. Usually with popular support (at first), they have enacted laws or policies which seemed to them and their supporters to be mild and common-sense attempts to protect the students' right to know there is more than one side to the story. This was almost always done against the advice of leading creationists, who knew what tactics the evolutionists would use to turn the attempt into another media and legal triumph for evolutionism.

All attempts to secure legal protection for anything other than evolutionism are doomed to this fate as long as: 1) The courts continue to hold an extreme interpretation of the so-called establishment clause of the First Amendment, 2) Major science organizations continue to define science in a way that has turned it into applied atheism (for all practical purposes), disguised by the lame excuse that scientists can still believe in a god – while working to explain everything in the universe as if there weren't any, and 3) The major media are ready and willing to portray the slightest move away from the monopoly of evolutionism as an attempt to force evil, ignorant fundamentalism on students.

Given those three rings to perform in, it is no wonder that evolutionists have turned almost every relevant court case into a circus. The lawyers for the defense generally try to keep the case limited, and almost always seem to be surprised and shocked at the way the judge is happy to let the lawyers for the evolutionists (usually including members of the powerful ACLU) bring in other issues, especially the establishment clause, and thus also bring in experts who otherwise would have no say in the matter. It's a kangaroo court once the judge goes along with points 1 and 2 of the previous paragraph, and the public support for the creationists vanishes as the media harps on the legal expenses, provides a stage for the evolutionists to parade their authority, and trots out a clown brigade of religious leaders bending over backwards to kiss the idol of the one belief that atheists love best.

The evolutionists pat each other on the back and claim their victories are due to nothing but the scientific strength of their theory. Unfortunately, their victories in the courts come at a dear price for true science, for evolutionism is being protected from any alternatives that might provide the truly rigorous testing and differing viewpoints that real science requires. Perhaps the leading evolutionists are fully aware that much of what they believe and want to teach exclusively in science classes is a matter of personal preference and philosophy rather than science. At any rate, it should be the greatest shame of all evolutionists that they are censoring debate and stifling alternative investigations rather than welcoming them.

Section VI: Conclusion – History Future

In contrast to the claims of complete victory by Niles Eldredge, Richard Dawkins, and other evolutionists, a number of creationists and Intelligent Design advocates have predicted the downfall of the evolutionary establishment sometime soon. It seems to me that the battle could rage for years. Theoretically it might never end. (It will probably be brought to a halt by events prophesied in the Bible. That may sound incredible to those who don't believe in the Bible, but the observations of things which match or prepare the way for events described in prophecies of the last days continue to increase.)

If the debate were a matter of pure science, then evolutionists might have proven their case long ago. Or rather, accepted that there are major aspects of their theory which can never be scientifically proven, and they would have made room for creation science. If it were a simple matter of everything in science supporting evolutionism and nothing but religion on the creationist's side, then the creationists, for the most part, no matter how "fundamentalist," might long ago have joined the mainstream religious groups and somehow adapted their beliefs to theistic evolutionism. Perhaps it would be called Biblical evolutionism. Certainly there would be far fewer creationists. At one time, the belief that the Earth is billions of years old was accepted by some of the most fundamentalist believers.

What keeps the issue from being resolved is its essentially philosophical nature. In other areas of science, the ability to demonstrate things by repeated observations and experiments even doubters can replicate leaves no room for reasonable doubt. In religion, appeal to the sacred texts can settle disagreements among those who accept the authority of the texts and the exegesis. Ongoing disagreements that used to lead to war now just lead to new denominations or smaller groups – live and let live. In the Christian realm, the active work of God in creating everything in the universe in a short time merely thousands of years ago was not doubted for over a thousand years. It was only when people began choosing human reason and philosophy as the final authority in all matters that ideas about the Earth being unimaginably old and living things changing wildly over generations began to be thought reasonable.

Evolutionism, however, can't be scientifically observed and demonstrated, and while it's vital for atheism, it's also been made to fit Christianity. So it's theoretically possible the two sides could remain deadlocked, each side unable to land a finishing blow because they are both ultimately based on the personal choices of individuals. Young-Earth creationists choose to believe that God provided Man with a revelation in the Bible that is trustworthy in a simple and direct way from start to finish. Evolutionists choose to believe that scientific interpretations of nature can and must reveal a completely natural history of the universe. There are all sorts of in-between views. The Intelligent Design position is that scientific observations are most consistent with some intelligent influence being required for the formation of life, but that position cannot resolve the question of whether God created life directly or merely created the universe in just the right way so that life would naturally evolve. It seems that the majority of people don't want to risk standing too far to one side or the other, but try to combine creation and evolution with vague platitudes such as "God could have created using evolution."

What might (theoretically) resolve the issue one way or another? One trend indicates that evolutionism might triumph, for all practical purposes, by having creationism outlawed. In Canada and Europe, strong laws have been passed to back up the evolutionists' monopoly on education, in some cases seeking to control what is taught in parochial schools as well as in the government-operated ones. The legal system in America is being bent toward the same end. Will it continue until creationists are forced to keep their views entirely to themselves? Whenever evolutionists defend the monopoly of evolutionism in science classes, they attempt to mollify the voting majority (who favor teaching creationism also) by claiming it would be perfectly acceptable to discuss creation in other classes – but such options are almost never seriously considered. On the rare occasions such subjects are proposed for other classes, the evolutionists then find reasons to oppose them, too, in spite of their earlier suggestions. A few extremists have already suggested it should be illegal for parents to teach their own children to believe in a Creator God.

Theoretically, evolutionists could present such strong scientific evidence for their beliefs that all but the hardiest creationists would be converted. The evolutionists believe they already have done so, but the growth of the creationist and ID organizations suggests that they're only fooling themselves. Until scientists have created life in the laboratory from scratch, and forced or aided a population of organisms through a series of mutations that give it a new, dynamically complex feature, we can't even make a good estimate of what it would take for such things to happen in nature.

The trend seen in the growth in number of creationist and ID organizations suggests that, theoretically, evolutionism could become a minority position simply by a wave of popular sentiment. However, the evolutionists' hold over education, science organizations, media, and courts gives them tremendous power to indoctrinate the next generation and to quell dissent. Yet again, it is just possible that their increasingly extreme application of this power will create a popular backlash.

Theoretically, scientists would grow tired of a science which shows such little advancement and few, if any, practical applications. In practice, scientists continue to be excited by such trivial things as the possibility that a variation in an insect population's color might lead to a separate population which could be called a new species. It is also easy to overlook lack of real progress and practical benefits to others when one's career is being advanced and one's personal finances are growing in response to repeating the party line.

Theoretically, the increasingly elaborate explanations, missing data, and difficulties of evolutionism, along with the mysteries and sheer weirdness of advanced physics and astrophysics, could make a majority of scientists realize that it is not unreasonable to recognize some supernatural activity probably affected our universe at some time. As more research is done using various models of Flood geology, and mainstream geologists find more evidence for ever larger catastrophic events, the links between geology and evolutionism that were forged under the strictest uniformitarian doctrine could weaken. In practice, however, it seems no amount of "negative evidence," for any length of time, is sufficient to make people doubt, especially when they have enough others sharing and supporting their faith. Likewise, no matter how well a new alternative model fits the data, it won't even be considered by those unwilling to think outside the fundamental axioms of the favored philosophy.

The final failure of evolutionism, however, may be its failure to replace the biblical Christian belief with a satisfying alternative story. There are signs that the void left by the suppression of traditional Christianity and creation from public venues will not be filled with the purely naturalistic "scientific" view, but with new forms of mysticism and religious compromise.

The wave of "modernism" which arose in mutual support with evolutionism has had rivals other than fundamentalist Christianity and creationism almost from the start. One of the recent views is called Postmodernism, which is antithetical to naturalism (materialism), reductionist science, and atheism, and is rejected by Humanists and fundamentalist Christians alike. Its roots go back to the Luddite and Romantic movements that formed as a backlash to the Industrial Revolution and Rationalism, but took its current form in the 1960s. It is fundamentally evolutionary, but it proposes the existence of a mystical purpose and intelligence in nature that is contrary to the pure naturalism of the scientific establishment. The mysticism and spiritual nature of the view is also contrary to the humanistic progressivism that was held by several leading evolutionists of the past. Yet these ideas sometimes find a degree of acceptance within the establishment which is vigorously denied to ID theory, although there are similarities and even some overlap. Also overlapping or fully within postmodernism are various New Age ideas (modern variations on Asian and primitive mystic traditions), and vaguely deistic ideas like George Lucas's "Force," etc. Popular works have tied a global ecology theory to the goddess Gaia for which it was named. Other books have compared concepts in advanced physics to teachings of mystical Eastern religions.

Evolutionists may be "fighting last year's war," chasing the phantom of yesterday's creationism and attacking the straw-man arguments they created themselves to misrepresent the creationists' positions. Meanwhile, the real threat to science may come from groups that creationists also oppose.

Remember, it is only in matters of the distant past, particularly in the origins of things, that creationists greatly disagree with evolutionists. All of the first "natural philosophers" and many great scientists were creationists. Most knowledgeable creationists understand and accept the rigorous, technical, scientific theory of evolution. But there are philosophical and religious schools of thought which teach truly anti-scientific doctrines. They argue against or propose alternatives to the objective nature of reality, the systematic analysis of nature, the sequential and non-repeating nature of time, and the reliability of natural laws. These ideas are creeping into scientific journals and also into ostensibly Christian churches. Such vague, mystical, and emotional views do not encourage the rigorous intellectual efforts needed for scientific research. Perhaps the debate will be lost in a popular fad of theistic evolutionism or deistic creationism which simply doesn't see any value in attempting to figure out any of the difficult details.

Although evolutionists see belief in creation as an outdated religious faith and creationists see evolutionism as a philosophy that has proven to be a failure, it should still be possible to interact without rancor and ill-will. There can be more in common than most people realize. Creationists can accept that organisms change over time and new species become isolated from old populations. Many people who consider themselves evolutionists concede that God may have had a hand in the origin of life or the design of a universe that would foster intelligent life. Remember, there is a complete spectrum of views. Even within the camps at the extremes there are variations and differences of opinion. A couple of YEC theories allow for the universe at large to be billions of years old. Some evolutionists have entertained or even promoted the idea that there is some purpose or direction to the course of evolution, or intelligence at work in nature. Comparing and contrasting ideas, mixing and matching models, and trying to see things from different viewpoints – these are all proven ways to stimulate new ideas and improve old ones. While evolutionists don't want to give up their monopoly, and young-Earth creationists would like everyone to see things our way, perhaps the best situation we can hope for is an increase in civilized and constructive discussions. Cooperation between evolutionists and creationists, or at least honestly and calmly considering each other's arguments, would only sharpen and clarify everyone's thinking.

Chapter 12: It All Goes Back to Fundamental Decisions

Introduction: Does debating details do diddly?

And so the debate rages on, largely because the two sides are playing two different games, and there is much confusion about which games they are playing. It's likely there will be people, including practicing scientists, holding both views until someone invents an actual time machine or God intervenes. The difference between the views is not between religion and science, but between the original philosophy of science which focused on things that are demonstrable or repeatably observable (leaving unique events of the distant past to written records and especially the Bible), and the modern philosophy of science which seeks natural explanations for everything that ever existed. The foundational assumption (which cannot be tested) is that natural forces have always operated everywhere as they do today, without any intelligent design, plan, or purpose involved. Modern scientific practice also tends to treat whatever seems to be the best such explanation as fact, or at least commonly presents it that way to the general public.

Thus, it is not merely religious stubbornness that keeps creationists from joining the evolutionary fold, but a strong distrust of supposedly scientific beliefs which cannot be repeatedly demonstrated to be reliable. No matter how evolutionists pile up fossils and observations of slight changes, the idea that microbes (much less raw minerals) evolved into men can never have the support that is available to Newton's mathematical descriptions of motion, momentum, and the effects of gravity. Likewise, the evolutionists' dogma that everything in the universe can and must have a natural explanation means that no matter how creationists point to gaps, holes, shoddy patches, kludges, and fudging, evolutionists will continue to have faith that the gaps will be filled, the explanations will be improved, or better (but still natural) explanations will be devised.

Ultimately, then, it comes down to a personal preference for one view of the world or another, depending on a faith in something beyond the established facts upon which everyone can agree. For some creationists it is enough to say "It's in the Bible and that settles it." Likewise some evolutionists have complete faith that natural forces alone account for everything. It is no help that still others in this postmodern age hold that there are different but equally valid truths for different people. So is there any point in further research and debating? Has no progress been made since Darwin's day? Is there very much that practically everyone on both sides agree on?

Perhaps not so many people are set in their ways, beyond hope of change, as it may seem. Certainly many impressionable young people have been indoctrinated to believe in evolutionism by their public school or college education, but likewise many people have later taken a hard look at what they were taught and found it wanting. A preacher might be moved to accept evolution after learning about a new kind of fossil.

Likewise one fossil may start an evolutionist science teacher on the road to rejecting evolution.

The general positions and irreconcilable differences remain, but scientific discoveries and observations have changed the details of what both creationists and evolutionists believe and argue about since Darwin's day. Today we have a practical understanding of heredity and many aspects of biological reproduction. The true nature of genetics was totally unknown in Darwin's day. So Darwin's notions of inheritable variations are now irrelevant and the debate over what biological variations can lead to has some hard facts involved.

We have had over 100 years more to observe variations within artificial selection programs and in nature. Everything from the breeding of exotic aquarium fish to cattle breeding to crop hybridization shows that rapid increase, decrease, or loss of existing traits is possible but reaches a limit, and no new dynamically complex features have appeared. Hybridization experiments (and accidents) suggest that supposedly long-separated populations are still within the same basic kind. Hybrids can be superior to their parent species, consistent with species being isolated fragments of a larger, more robust gene pool. All this totally undermines Darwin's main argument. Variation in domestic species is still sometimes presented as strong evidence for evolutionism, but while it shows the sort of change which the scientific theory predicts happens in the wild, it is only assumed that such changes could somehow produce all living things from microbes.

We have done numerous studies in natural selection, and all of them show that it is capable of shifting the ratio of variations in the gene pool, but that's about all. Mutations almost always cause the loss of something, make something worse, repeat parts (almost always useless or harmful), or make physically trivial changes such as color variations. The extremely rare beneficial changes are no different, they just happen to be useful in a certain situation. Some beneficial changes seem to be part of a designed adaptability involving a fixed set or range of variations.

We also have had more time to look in more places for fossils. The results are championed by evolutionists in that more intermediate forms have been discovered, supposedly filling in the gaps. On the other hand, creationists point to non-transitional aspects of the intermediate forms, and to the rejection of forms that were once thought intermediate but are now recognized as unique and too specialized to be transitional. Worse yet, some were rejected because they were badly misinterpreted or, in the case of Piltdown Man, a hoax. Above all, Darwin's nightmare, the Cambrian explosion, is still something evolutionists have to sweep under the rug with excuses. Also, pterosaurs, bats, and other very distinctive forms of life still have no significantly intermediate "ancestral" fossils. There are new distinct and unique fossils discovered since Darwin's time. The common pattern of "sudden appearance" of many new forms at various levels remains, a major departure from expectations and genetic extrapolations. Fossils that appear much like living forms continue to be discovered and assigned earlier dates. A relatively few possible intermediate forms while vast numbers show whole new sets of life appearing together, and hundreds of millions of years of apparent stasis, is far from the evolutionist's vision of life constantly and very gradually evolving without discernible limits.

So, the debate continues with generally the same rhetorical volleys being hurled across the same fundamental philosophical division, but changing and becoming increasingly complex (when the participants are informed, civil, and willing) in the details. Some leading evolutionists (who are parroted by the mass media) seem to believe there's no need to go into details, because anybody who doubts Evolution must be clinging to faith in religion that is contrary to science in general. They are unwilling to see their own evolutionism depends on faith that remains unsupported, or is incompatible with important facts.

Section I: Faith and the Great Unknowns of Evolutionism

No matter how much evolutionists want to believe that their position is the only one possible given the available evidence, or that they are only accepting facts, they are blinded to other possibilities by their conception of science. Anyone who (in effect) professes faith that everything that ever happened has a perfectly natural, intelligence-free explanation must have faith in three general explanations for how life came to be as we know it, regardless of any lack of evidence, or even contrary evidence.

1) The first is that complex living things must have come from simpler living things. This is the first step from the technical, scientific theory of evolution to Evolutionism. This first point is required because it is not natural for complex things to pop into existence suddenly, and nothing natural has been around forever. So the only possible natural explanation for living things is that they came from something simpler.

2) The second explanation follows from the first: if complex life came from simple life, then all the different forms of life must have come from a few very simple ancestors, or just one. This is the general view of biological Evolutionism that is commonly confused with the scientific theory.

3) The third explanation is that the first simple form (or forms) must have evolved by purely natural, unguided processes from non-living matter. This is technically not part of biological evolution, but it is the only possibility consistent with the view that requires the first two.

These remain, ultimately, a matter of faith, because no one has ever observed a population of simple organisms producing a new form that has an increase in systematically organized dynamic complexity. In other areas of science, this sort of shortcoming would keep an idea in the category of speculation or perhaps hypothesis. In evolutionism, it is covered up by appeals to observations of other kinds of changes and speculative explanations for similarities and differences in different living things. So, while the debate can become very detailed and complex (interesting and informative as well), it still comes down to this: anyone who ascribes to the current view of science has no choice but to have faith in an evolutionary explanation, but anyone who believes there could have been supernatural intervention in the past and holds science to its original method will find any of them unsatisfactory.

For example: To an evolutionist, a mutation which changes the color of a hatch of fish which then avoid mating with fish of the original color supports his faith that fish evolved from an invertebrate – little changes must accumulate to produce large changes. To the rest of us, it just means that fish can evolve into species with different colors – changes in color are different from changes that would produce new limbs or organs. Likewise, to someone who doubts evolution, observations of the kinds of changes produced by mutations is a strong argument for limited variability of living things. To an evolutionist, change is change and if mutations are the only heritable source of change, then they must somehow have produced all biological variations.

The evolutionist may point out that the rate of mutation is sufficient to account for the number of genetic differences, and the evolution doubter can point out that the rate doesn't take into account degenerative changes, reversals, and other changes which do not advance the population toward anything more complex. The evolutionist can respond with examples of beneficial mutations. The skeptic can point out that the beneficial changes are not due to an increase in complexity. Then the debate can get into the definitions of "new" and "complexity." Often, if the two sides don't recognize that they're arguing from two separate frameworks with different standards of evidence, different rules of procedure, and ultimately two different faiths, the debate breaks down into charges of intellectual dishonesty of one sort or another. This is easy to do because creationists want to be considered as scientific as evolutionists although they don't believe science can reveal as much about reality, and evolutionists cannot admit that so much of their "science" depends on faith.

The same chasing around the bush is inevitable in arguments over the second explanation of life. To an evolutionist, just as any observed changes imply that natural change can account for all variations, so too almost any observed similarity between different living things can be taken as evidence that they had a common ancestor. This area of debate may be even more complicated and messy than the first. Of course we observe similarities and variations in ancestors and descendants. On the other hand, we also know of cases in which certain similarities are not caused by inheritance from a common ancestor. Evolutionists chalk these up to parallel or convergent evolution. An apparently vestigial organ in one form of life may be evidence for common ancestry with another that has the fully-functioning version, but it also might result from degeneration of a similar organ in a separate ancestor. At best, vestigial organs show possible common descent from a more complex ancestor, not a more primitive (simpler) one. Studies of similarities and differences in DNA have produced some interesting results, but nothing consistent beyond what might be expected from similarities in whole bodies. There have even been surprises in which molecular analysis caused major revisions in the supposed relationships of similar organisms. Evolutionists sometimes argue among themselves over specific cases of possible common ancestry, and various creationists have different ideas about the limits of variation and common descent.

Unfortunately, this mutual uncertainty has seldom if ever helped to tone down the bombast or venom in a debate. Ultimately, the consistent evolutionist must go all the way to one or at most a handful of common ancestors, and Biblical creationists can't accept a mere handful of simple forms as the original created kinds without badly stretching the "interpretation" of the Bible's creation account. Few seem interested in exploring together the common ground, things like observed hybridization and recorded common ancestry. The only group that seems happy to explore common ancestry from one possibility to the other is the ID movement, which includes at the extremes of its "big tent" approach some YECs and some who accept everything about the standard theory of evolution except the belief that it all happened without any intelligent guidance or plan. Evolutionists might be surprised at just how much natural variation creationists are willing to accept. If they ever notice, they don't dare say anything positive about it, to avoid accusations of being "closet creationists." Consistent evolutionists cannot show any interest in any possibility of supernatural or intelligent input in the history of life.

Perhaps the greatest test of an evolutionist's faith is in believing that a living thing can and did (at least once) come into existence through nothing more than an accidental combination of chemicals and energy. Nothing at all close to this is seen in nature. Evolutionists cover up the fact that this is a matter of faith with – well, not much in this case. They might try to avoid the issue by allowing for divine interference at this point, but that is inconsistent. They might argue that the origin of life is not a part of the theory of evolution, but that is ducking the issue, putting off the inevitable. Remember, a great deal of fuss was made over the first experiment to produce some basic organic chemicals under conditions that were thought to replicate the natural environment of the early Earth. Likewise biology texts often discuss the origin of life in their sections dealing with evolution. But the excitement over the Miller-Urey experiment turned out to be a false dawn. After decades of trying, no highly intelligent researcher has managed to create a new living thing, although it's an event which supposedly happened as an ordinary chance conjunction of raw chemicals and energy.

Not surprisingly, there's not as much debate on this point. The evolutionists who care to get into it have to make lame appeals to what little progress has been made, what future research might produce, or the notion that an infinite number of universes has existed, and we just happen to be in a lucky one. A lot of the debate on the issue is between evolutionists who have different stories about how life might have started. If one of them were ever to be demonstrated, then there would at least be something solid for evolutionists to put their faith in, but for now they have faith in spite of all the observations that living things end up as non-living matter, and non-living matter has never been observed to spontaneously become alive. Even a disassembled watch will not form a whole one again without the application of intelligence as well as energy.

Section II. Facts and Fantasies

Subsection A: Interpretation of the Geological Record & Fossils

So, in spite of the way it is commonly presented, the debate is not a matter of science versus faith, but faith in natural processes versus faith that God clearly revealed the true history of the creation. This hidden error has caused and is hidden by the complexity of the debate, with its spectrum of different positions, basic but unstated differences in philosophies, theological arguments from all sides, misunderstandings all around, and the flurry of all the scientific terms and data. The strongest creation position holds that science went too far when it began pontificating about the distant past. The debate wouldn't be so heated if scientists had always made it clear that such statements are provisional and speculative. Evolutionists will indeed affirm that their conclusions are subject to revision, but when asked to debate the issue on purely scientific grounds, it seems they always treat their "scientific" beliefs as facts, and use them to attack Biblical or theological issues. The ID movement tries to leave faith out of it entirely, while allowing leeway outside of their research for those who have faith, but the evolutionary establishment vigorously opposes any departure from strict naturalistic philosophy.

Still, there is a great potential for agreement in matters that are important and practical, namely the sort of slight variations and immediate common ancestry that is observable and has been observed. As far as things such as variations in domestic breeds, production of modern crops by hybridization, microbes exhibiting resistance to antibiotics, natural selection shifting gene ratios within populations, and even speciation by reproductive isolation, creationists almost universally have no objections.

There is one area of debate where there is 1) a lot of evidence to argue over, 2) there is real debate and not just one side claiming the other doesn't accept the evidence, 3) it's not just Young-Earth Creationists who doubt the mainstream story, and 4) the evidence (as presented by major public sources) appears at first to strongly favor Evolutionism. This area is called the fossil record.

What seems to impress most people more than anything else is the overall pattern. Most creationists will argue that there are few if any places on Earth where the entire sequence of rock layers can be found in the exact order specified by the accepted evolutionary history. That's at least debatable, but what's undeniable is that different layers of rocks have different sets of fossils in them. The core of this part of the debate is thus a question of what the fossil record actually represents.

To evolutionists, the fossil record has always shown the history of life evolving from simple forms to increasingly more advanced forms, most notably from fish to amphibians and on through reptiles and then various mammals, including humans. This is perhaps the most iconic image of evolutionism: fish-amphibian-reptile-rodent-monkey-ape-human. While there have been some debates within the evolutionary camp, they have all fallen well within the broad Darwinian boundaries. The debates tend to be over whether evolution proceeded smoothly or in fits and stops, how large the steps may be, whether or not it was guided or driven by natural forces, is the history largely inevitable or might it have been very different, does it have some direction or not. It is enough for many evolutionists and many old-Earth creationists that the geological pattern doesn't appear to represent what they suppose a global Flood would produce – as if we could imagine everything about such a huge, complex, unique and to some extent supernatural event.

Those who look closely and with a scientifically skeptical eye at the fossil record, however, see that it is not what the theory predicts. Here is another area where there can be a little agreement, although very few evolutionists dare to stray far from the established dogmas, and creationists have many different idea about it. There is a wide range of theories and differences of opinion about just how much of the fossil record is due to the Flood even within the young-Earth organizations. On the other hand, some evolutionists agree that the pattern of fossils is not consistent with the slow and gradual evolution predicted by the standard theory. Evolutionary Intelligent Design theorists and progressive creationists suspect some intelligent adjustments were required at some points, and some theistic evolutionists see stages of a divine plan.

At this point, then, the criticisms over using the fossil record as support for totally natural, slow and gradual evolution are somewhat orphaned in the sense of not being accompanied by any strong alternatives. For many of the middle positions, there may not be any hope of positive evidence for an alternative explanation. The young-Earth, global catastrophic flood position has been maintained (more or less) by a few geologists now and then. As they have been few in number and have little or no funding, it's not surprising that they haven't yet agreed on a consistent interpretation of the vast amounts of geologic data. Still, they have done much in the way of pointing out features that are inconsistent with the standard dogmas of evolutionary geology. There are also a number of general observations and considerations which can show the fossil record to be at least as consistent with a global Flood as an evolutionary history.

The appearance of a wide variety of new forms from one geological division to the next, the appearance in many layers of similar fossils with only minor changes (and even fossils of things such as tree trunks that each pass through multiple layers), and the living forms which are virtually identical to forms supposedly tens or hundreds of millions of years old all raise questions about the standard interpretation of the fossil record and are consistent with a recent, rapid process.

The general pattern which supposedly supports evolution can also be interpreted as (for the most part) representing altitude-associated ecological zones. An early, if not the first, ecological zonation model of the fossil record was put forth by Harold Willard Clark (1891-1986). "He was a professor of biology and geology at Pacific Union College, Angwin, California. Author of The New Diluvialism in 1946," he was an Adventist like George Macready Price, but "A rift developed between Dr. Clark and George McCready Price who did not accept order in the fossil record for many years." ("Ecological Zonation: Theory, Problems, and Perspectives" by Elaine Kennedy, Geoscience Research Institute https://adventistbiblicalresearch.org/sites/default/files/pdf/Ecological%20Zonation.pdf)

The following is just a simple suggestion, and of course something produced by a catastrophic destruction of a complex biosphere would have many other factors involved. Factors that would complicate the picture include variations in mobility of life forms, variations in postmortem flotation, the effects of currents, tides, back-flows and tsunamis, similarities in kinds that were created separately for different environments, and changes in life forms as adaptations to environmental conditions. However, as a general rule of first appearance, I believe it works quite well.

Thus, the lowest fossils are kinds that lived on the ocean bottom or were relatively poor swimmers. The first definite fish are higher up, and the first land plants and amphibians higher still. Reptiles, which are well designed for very warm, dryer terrain appear next. Meanwhile, the early fish which tended to have heavy armor or other features (or lack of them) limiting their speed and maneuverability fade out and lighter, faster-swimming species show up. Is this an improvement or merely better adaptation for surviving the extreme conditions of the Flood? The mammals and birds, especially those most like living ones, have the best designs for living at high altitudes and surviving changes in temperature. They also have features (thick coats of oily feathers or hair, large lung capacity, layers of fat, etc.) which cause their corpses to float longer than those of reptiles.

As that last point suggests, the various sorting effects that water can have would also play a part in creating the fossil pattern. It would also affect the pattern of the fish fossils, as the armored ones and other "primitive" forms would tend to sink right to the bottom, while those with swim bladders would float for some time after death. One of the main features of the record, the complete extinctions of many forms of life, seems more consistent with the Biblical Flood account than it is with evolution. It's true that there are a number of features which do not seem to fit the Flood scenario, but considering that there are also problems with the evolutionary interpretation in spite of the great number of geologists with all the government and academic funding and equipment who have been working on it, that doesn't seem like such an impressive consideration.

Another aspect of the fossil record that many find very supportive of evolution is that of specific "transitional" fossils. Evolutionists have followed in Darwin's footsteps by making all sorts of excuses for why the fossil record isn't filled with transitional forms, ignoring that all but a few fossils are of known "terminal" or "crown" types, and playing up every possible intermediate or shared feature. Some have explained that they don't expect to find the actual missing links, but they continue to claim that the intermediate forms found show that the transition was possible and that "something like" the fossil organism, a "sister species," with intermediate features did exist and was the true transitional form.

It seems creationists, especially at the beginning of the modern creation science movement, made the job too easy for the evolutionists on a few points. According to a number of claims made (and which are still repeated by some creationists), not only did the transitional series never happen, it would be impossible for certain intermediate forms to exist, and so they would never be found. This idea was commonly expressed in catchy, popular-level sayings such as: (regarding the origin of mammals) "How is a reptile going to eat while dragging its jaw bones into its ear?" and (in response to the story that birds evolved from dinosaurs) "What good is half a wing?"

Such statements (and the detailed scientific arguments behind them) are misguided for at least two reasons:

1) They overlook the fact that a smooth or complete series of organisms had been posited long before evolutionists thought of it, in the philosophical-religious concept of the Great Chain of Being. While that extreme idea was misguided, obviously the existence of a great series of intermediate forms could be consistent with creation by a God of infinite wisdom and imagination.

2) They overlook the possibility of some intermediate forms being due to changes that are in the opposite direction from the evolutionary story. For example: If a kind or whole class of birds were created with more dinosaur-like features (such as teeth, claws on the wing "finger" bones, and long bones in the tail), one of them might have

a mutation destroying its ability to fly (perhaps degenerate forelimbs, creating "half a wing") but allowing it to survive as a flightless bird. If it managed to start a population of similar descendants, the fossils would have the same appearance as a bird-like dinosaur evolving into a bird. And of course there is the probability that God created some flightless dinosaur-like birds in the first place, just as some species of extant flightless birds were probably created pretty much as they are.

At any rate, intermediate forms were found and the evolutionists rejoiced. Many of them considered their case "proven," and they expected the creation science movement to dry up and blow away. However, a few fossils with some intermediate features does not show that an entire evolutionary transition took place. The objection that the fossil record does not show an evolutionary pattern remains. Of the relatively very few fossils with some intermediate features, there are even fewer cases where the dates assigned to them are lined up in the evolutionary sequence. The expected general appearance of constant, gradual change is not supported by the fossils. Evolutionists still have to deal with many large gaps in the fossil record (exactly where or "when" the true transitional forms would have to be) and appeal to "ghost lineages" to explain away the "reverse" order of fossils.

Finally, even within the evolutionary playbook, there are concepts and cases which call into question the certainty with which intermediate forms are declared to be (or illustrate) evolutionary transitions. There are many cases where evolutionists have concluded that shared or similar features are due to parallel or convergent evolution. If such similarities that are not due to common ancestry or an ancestor-descendant relationship exist within the evolutionary view, can mere appearance of similar features in a relatively few fossils be strong evidence that evolution produced all life from microbes?

Subsection B: Connecting the Dots When They Seem Close Enough

Let's look at what might be considered the best cases of fossils which appear to show some important intermediate steps in evolutionary transitions. We'll start with those in what are generally the lower rock strata and work our way up.

The first major case that evolutionists like to present is that of the supposed evolution of some extinct fish into extinct forms of amphibians. For a long time, they didn't have much to show, but a number of reports in the 1990s and later gave them confidence in saying the gap between fish and amphibians was filling in or had been filled, and there should be no doubt that the latter evolved from the former.

One neat and pretty version of the story goes something like this: One group of fishes (including *Eusthenopteron*, *Elpistostege*, and *Pandericthys*) had evolved leg-like paddles for their fins, complete with bones matching the three main bones in the arms and legs of land animals. Competition and predation in the open water was fierce, so some of these fish (such as *Tiktaalik*, which had extra flexibility in its neck and bony fin supports) moved into shallow, swampy waters to get away from all that. The water was murky, low in oxygen, and choked with plants, so the fishes which happened to evolve primitive lungs (in addition to gills) for gulping air, stronger limbs for pushing up from the bottom to do so, and fingers for hanging onto plants, had an advantage.

These were transitional forms such as *Acanthostega*. As these lungs and limbs evolved further, it allowed the new amphibians such as *Ichthyostega* to get further away from the dangers in the water and also to feed on invertebrates that had earlier colonized the land. This allowed later amphibians to rapidly evolve into many different forms.

It does look good, with several stages and intermediate steps. For those with faith in the power of evolution, the case can be considered closed. Why shouldn't they feel satisfied with a few possibilities and leave it at that? For other people, there are several good reasons for doubting.

The first thing to consider is that if we only had living animals to go by, evolutionists would consider extant lungfish and amphibians to be (or to illustrate the possibility of) links between fish and animals adapted to dry land – the theory demands some such series occurred. Thus, the strength of an evolutionary argument can't be judged by the confidence of its proponents. In other areas of science, an idea isn't considered truly tested and accepted until doubters are convinced, but with evolutionism, there's no real alternative, only variations, so all doubters are said to be unscientific, and thereafter ignored.

The first detail we can be skeptical about is the first step, that of fish evolving limb-like fins. There certainly are such paddle-finned fish, but there's no evidence that such paddles gradually evolved. And how and why would they? Bony-finned fish fossils appear fully-formed, along with cartilaginous fish and bony fish with soft fins, in the same formations. These layers are the first (i.e. "oldest" or usually lowest) in which fish fossils are common. Aside from some rarer and usually very fragmentary traces, it's as if all three major groups of fish appeared at once.

The fossil fishes known as coelacanths had been predicted to "walk" on lake or sea bottoms, or even to sometimes crawl up on the shore. They were sometimes called "Old Fourlegs." They are dated about as old as any other bony-finned fish, and were supposed to have died out long after the first amphibians appeared, but still 65 million years ago. Then it was discovered that some were still alive – and living deep in the ocean.

The geologists and paleontologists had concluded that there had been great droughts all over the world (or much of it) during the Devonian. They believed and taught that these frequent severe dry spells had forced fish to crawl out of pools in search of more water, thus fostering the evolution of amphibians. Even long after the living species of coelacanth had been discovered and found to lack any tendency to try walking on land, paleontologists held and promoted this view. Finally, though, geologists concluded that they had misinterpreted the record in the rocks and that the Devonian hadn't been subjected to extraordinary droughts after all. The evolutionists did a complete re-write of what had once been taught to students as if it were a fact, and the story became that some fish used their fins to pull themselves through plant roots and stems in swamps, or to push off the bottom near shore, in order to breathe air when the water got too stale in the swamp.

So the presence of bony fins on fish in apparently shallow water deposits is no guarantee that they were using those fins like feet at all, let alone evolving into amphibians. We can't even be sure of our interpretation of the environment the creatures normally lived in. Some fish move into the shallows or up rivers and streams to spawn. Dead fish can be carried into the shallows or onto shore by currents, tides, and storms. While there are fish with fleshy fins and lungs, there's no known fish with bony fins and lungs.

There's also no real connection between the more amphibian-like fish such as Panderichthys or Tiktaalik and the amphibians such as Acanthostega, only technical similarities such as in some skull bones. While Acanthostega does have skeletal traces that appear to have served as supports for gills, it has no gill covers and there are living amphibians (axolotls) with external gills that are not believed to be inherited from fish ancestors. It is not at all certain that Acanthostega's gills (if indeed it had them) were retained from more fishy ancestors.

Most distinctive, however, is Acanthostega's digits (fingers and toes) that even a child could recognize, and definite (if weak) skeletal structures connecting both the forelimbs and hind legs with the backbone. All bony-finned fish have fins on the ends of their paddle-limbs and none of them has the connections between their limbs and backbones. In a line-up of their forelimbs, Acanthostega's digits stand out like a sore thumb (pun intended). For that matter, the line-up is generally based on similarities in other parts of the animals involved and/or their assigned dates, and there's no sign of progression in the limb bones themselves. Show the set to average citizens and ask them to arrange them in some order, and you'd likely get a number of different line-ups, quite likely with none of them more frequently matching the official evolutionary lineage. However, everyone would place Acanthostega's foot at the end, or even set it off by itself as not belonging with the others.

If you press the evolutionists on the point, they will explain that it isn't actually an evolutionary lineage, merely a cladistic phylogeny. In other words, they know that perhaps none of their samples represent a member of a population that descended from or gave rise to one of the others. Each case is only supposed to be the closest we have to a representative of such a population. This is true of almost every single "transitional" fossil that evolutionists offer up.

Finally, there is something about the case of *Ichthyostega* which (for me, at least) gives reason to doubt any claims by evolutionists about the strength of their evidence for major evolutionary transitions. For decades, *Ichthyostega* was the prime example of the transition between fish and land animals. Features such as its "fish-like" tail and relatively weak limbs were proudly presented, and little, if any, indication was given to the public that there might be features that would not be found in the line between the fish and amphibians. Then *Acanthostega* was found and described, providing a new and arguably better possible transitional form. Not very long after, new research reports revealed that *Ichthyostega* had a couple of significant unique features. It had an ear structure unlike anything else's, and its backbone was built to flex up and down, like an otter's, rather than a fish's.

The evolutionists cited a good reason or two for why these clearly non-transitional features hadn't been better studied and reported to the public earlier, so they can't be charged fairly with a willful conspiracy to keep us in the dark. On the other hand, it still leaves us with a striking example of evolutionists telling us that a fossil was a great example of a transitional form, when it turned out to be far too unusual or specialized to fill that role. If they could be so wrong in that case, why should we not maintain our doubts about others? Especially since this is hardly the only such case.

The next group or "series" of more-or-less transitional forms that evolutionists like to present is the reptile-to-mammal story. The main problem here is that some of the most significant changes that would have to take place involve soft tissues such as mammary glands and reproductive organs, which almost never get fossilized.

As with the supposed fish-to-amphibian transition, there are living animals which are said to be closer to the transitional forms than others. In this case they are the monotremes, which lay eggs and exude milk from patches rather than suckling their young. Yet there is once more an "on the other hand" here. Both of the living kinds of monotremes, the platypuses and the echidnas, are very specialized animals, with unique broad beaks (in platypuses) and extremely pointy snouts (in echidnas). They both have true hair (not long scales or scale-like hair) and, although they don't have breasts or nipples, they do produce milk. Thus, the evolutionary stories of how the most distinctive features of mammals came to be remain a matter of sheer speculation – faith in fantasy.

In this case, too, we find an example of a form that was once presented as a transitional that is now given a new position. At one time it was assumed that marsupials (kangaroos, opossums, etc.) were in a primitive, transitional stage between monotremes and placental animals (pretty much all the mammals that non-Australians are most familiar with). More recent studies have placed marsupials on a separate-but-equal branch of the evolutionary tree. Here again, how these two distinct and equally amazing systems of birth and infant care began can only be accounted for by faith in either divine creation or evolutionary speculation, not by scientific demonstration or hard evidence.

So it is not surprising that the debate has focused on some differences between the skeletons of mammals and reptiles. The most interest has been in just one or two aspects: which bones are involved in the jaw joint, and the differences in the tiny bones in the ear (only one in reptiles, three in mammals). The evolutionists argue that some bones in the jaw shifted and a couple of them moved into the ear and became the extra two bones there.

The strong point for the evolutionists is that there is a whole class of fossils with skeletal features that suggested the name, "mammal-like reptiles." Some types are so reptilian that they were commonly included in sets of "dinosaur" toys. Others are so similar to mammals in skeletal features that only a couple of technical points keep them from being classed as mammals. Once again, it's not too hard to imagine a transitional series and line up a number of fossils accordingly, while thinking within the scientific establishment's box that demands such explanations. But the case is not so compelling if we allow that the philosophy involved may be misleading.

For one thing, the fossils don't show a pattern of gradual progression. The mammal-like reptiles themselves appear quite suddenly with a number of unique features and in a wide variety. In spite of their general reptilian form, they are clearly distinct from other reptiles. Meanwhile, fossils with the uniquely mammalian trait of three ear bones have been dated to the same general period. The key fossils supposedly showing the change to three ear bones look a lot like the same bones of modern monotremes, suggesting that they were likewise uniquely specialized forms and not transitional. Another major consideration is that the connection between any of the Mesozoic mammal fossils and living mammals is not at all clear. Most modern mammals show up as distinct forms some time after the Mesozoic-Cenozoic boundary. All of these kinds could simply represent the panoply of diverse, separate, original created kinds or design themes.

Another star in the evolutionists' drama of life is the story of birds evolving from dinosaurs. There may still be some evolutionists who believe that birds evolved from some other form of reptile, but the more exciting "Birds are dinosaurs!" theory has won the popularity contest by a landslide. For a long time, the only significant intermediate form known was Archaeopteryx, but in recent years a number of new forms have been discovered, allowing evolutionists to present line-ups showing several stages of dinosaurs becoming increasingly birdlike. These are presented to the public as powerful evidence (or proof) of major evolutionary change.

Looking at the actual data, however, we again see that it does not line up in the way the evolutionists want to connect the dots. Looking for such things with a skeptical eye, especially regarding evolutionists' attempts to explain away what doesn't fit, makes it clear that in this area, scientists commonly let their desires drive their interpretations of the data. Worse yet, comparing the data with the neat-and-pretty picture presented in popular media makes it clear that the public is being purposely mislead.

Let's start with the prima donna of evolutionary "transitional forms," Archaeopteryx. Evolutionism got a big boost when the first fossil was discovered not long after Darwin's *Origin* came out. In many ways, it looks like a very small bipedal dinosaur, like a small Velociraptor. But it has a full set of flight feathers, giving it a beautiful pair of wings. The perfect half-dinosaur, half-bird transitional form? Let's look closer.

Some people have claimed there is evidence the feather imprints are fakes, and on the other hand, many creationists have pointed to the wings and other bird-like features and suggested that Archaeopteryx was, more or less, just another bird, if an odd one. I believe both of these views are mistaken, to put it mildly. There are a number of fossils, and the original is not the only one to show traces of feathers. "Archie" was intermediate in form, but it was definitely a "bird" in the old sense of an animal with a backbone that could fly. This may be something of an annoyance for the Birds Are Dinosaurs (BAD) group, because with all the new "transitional" fossils that have been found, hardly any of the very bird-like ones have been found in deposits dated before Archaeopteryx, and those few are also fully feathered or otherwise a lot like Archaeopteryx.

Velociraptor, Tyrannosaurus, and other "ancestral" types are all dated as being more recent. This pattern makes it look like some dinosaurs descended from dinosaur-like birds! According to at least one recent analysis, Archie should be classed as a dinosaur, which would fit this possibility. It also fits the possibility that God simply created some bird-type dinosaurs and some dinosaurs that shared some features with birds (related to being bipedal and having a high metabolism) but neither descended from nor were ancestors of anything that ever flew.

This has led to one of the clearest examples of scientists abandoning the fundamental method when data contradict their pet theories. Science generally proceeds much more cautiously and tentatively. This is on top of the fact that paleontology in general posits conclusions about things that have never been observed, nor have similar phenomena been observed. In this case, even the circumstantial data is treated in a way that favors evolutionary expectations. Everybody knows that amphibians and reptiles don't have feathers and don't fly, so it is assumed that feathers and flight gradually evolved, and if the fossils don't show that pattern, then the fossils must be assumed to be oddly preserved by chance.

However, in spite of efforts to downgrade Archie's bird-like features and smooth the way for the BAD theory, it's not hard to see a few contrary things: Archie is dated tens of millions of years older than dinosaurs like Velociraptor. It has well-formed wings with aerodynamic feathers. It has forelimbs that are long, and "finger" bones that are very long and very thin. In specimens where these bones aren't broken or awkwardly bent, the digits are always straight, indicating they were involved in supporting the wing and never used for grasping. Two of the digits are always close together in such examples, as if bound together by ligaments or softer tissues, and incapable of independent movement in life. The tail is long, but extremely thin and inflexible beyond the base, supporting aerodynamic feathers. It has a toe designed for grasping, unlike any dinosaur, although the BAD guys argue that it wasn't all the way in the back and facing forward as in modern birds.

It is argued that fuzzy, partially-feathered, or just superficially bird-like dinosaur fossils that are dated after Archie are all we've found of populations that had begun tens of millions of years earlier ("ghost lineages"). However, there are also fossils of birds (feathered, flying vertebrates) that were much more specialized and better designed for flight than Archie in the same strata as these "bird-like" dinosaur fossils. So how do we know they didn't occur earlier, also? There are traces supporting this possibility. There is a controversial fossil named *Protoavis*, and tracks that look very much like those of birds. Some of the best examples, and the most bird-like, are supposed to be over 200 million years old, more than 50 million years before Archie! There are also fossils of several different kinds of gliding reptiles that clearly weren't dinosaurs nor ancestors of birds. Why should the fuzzy, feathered, or otherwise especially birdlike dinosaur fossils only show up in rock formations given younger dates? Apart from evolutionary bias, there's no particular reason to believe that dinosaurs evolved into birds rather than believing that there were birds all along, and that the supposed "transitional forms" are just variations (including flightless types) with some dinosaur-like features.

To top all that off, there is this large problem with the idea that modern birds are dinosaurs: as with mammals, none of the transitional "dino-birds" or more advanced forms appear to have gradually evolved into the wide variety of living birds. Instead, the last of the Mesozoic birds tend to be oddly specialized aquatic birds, while modern birds first appear (with the possible exception of a few waterfowl fossils in the last of the Mesozoic sediments) after some extinct large flightless birds, clearly distinct from the extinct bird types and about as different from each other and identifiable as they are today. This variety of modern bird fossils, and the differences in genes in living birds, has caused researchers to conclude that (despite the lack of fossils) all 40 or so modern orders of birds were already formed back in the Cretaceous, alongside the last dinosaurs. That's a lot of evolving for which there aren't any significant candidates for direct ancestral forms. Once again, a supposedly shining example of what could be a major evolutionary series of transitional forms turns out to be a smoke-and-mirrors illusion supported more by the faith of Evolutionism than by the facts.

Another highly-touted case of fossils which could represent a true transitional series is that which supposedly shows how land mammals evolved into whales. It starts with mammals that show no traces of adaptation for living in the water except for one feature in the ears, plus dense bones and high-placed eyes that, with its long, thin legs, suggests it didn't swim, but possibly walked in shallow water up to its eyeballs. The next dot is some fossils that look like long-nosed sea lions or giant otters with short legs. Similar forms dated as more recent are larger and have proportionately shorter legs. Finally there are forms that definitely look like small whales as we know them today, with certain arguably primitive features.

Even here, evolutionists don't believe the fossils (in most cases) are of the actual transitional forms, but are merely illustrative of how the transitional stages would have looked. That point is usually lost in the enthusiastic descriptions of these fossils in the popular press. Further examination shows other gaps between the popular image and reality.

There has been (and may be still, or again) some argument among evolutionists over which group of animals the first "walking whales" evolved from. Different studies have suggested different groups, so that the apparently favored answer has changed back and forth a couple of times. One fossil skull was presented as belonging to something like a large sea otter, but later fossils showed the type was more like a coyote. It was said to show that whales evolved from an odd group of carnivores with hooves, but later research returned the ancestral line to an unknown creature that would also be an ancestor of hippos (and pigs, and other hoofed animals). Also keep in mind that living otters, sea lions, seals, and river dolphins could be lined up to give the appearance of a transitional lineage, although evolutionists do not consider them to represent such a line.

The early semi-aquatic forms would have depended on their large hind feet for their swimming propulsion, like sea lions and seals do – why would their descendants have changed to having no hind feet at all? It should be remembered that loss of parts is the sort of change that fits with the creation framework, but losing a key feature of the locomotion system is still a drastic development.

Once again, the lack of preservation of soft tissues leaves major questions unanswered. For example, the origin of whale flukes is a great mystery. Would gradual mutations produce them, starting with little tail-tip nubs? How much of a nub would be more of a help than a drag? If mutations could lead to flukes on whales, why don't we see protruding nubs on the tail tips of other animals? Of course, flukes wouldn't be helpful to anything that doesn't swim, but it seems reasonable to expect that if they gradually evolved from some such small beginning on whales, we should commonly see harmless nubs similar to the beginnings of flukes at various places on other animals. Why would it happen to happen only on whale ancestors?

Still, the whale series at least has a line of plausible-looking candidates that are assigned dates in the proper order (although the dates assigned to some are bunched together and there are noticeable gaps at other points in the line). Much of the story from the point of "archaic whales" onward might fit within the creation view. If such major series were a common feature of the available data, and no lines with striking gaps were known, then the case for evolutionism might be persuasive to those of us who aren't personally or philosophically committed to it. However, as this examination has shown, evolutionists usually have had to arrange the fossils to their liking and gloss over problems when presenting their "transitional lineages" to the public.

One case, the horse series, has fallen out of favor among knowledgeable evolutionists. The connection between Hyracotherium, aka Eohippus ("Dawn Horse") and the next step up is not at all clear. After that, there are many varieties of horses, but there's no clear line leading to modern horses. Even if there were, the kind of changes involved (increase in size, loss of toes, number of ribs, changes in proportions of teeth and other parts) also fit within the creation framework.

This leaves just one other major evolutionary transition that evolutionists have managed to impress the public with: the story of our own origins from some ape-like ancestor. It should be remembered (although evolutionists seem to think it should be forgotten or not considered significant) that this story has had many revisions, and has been plagued by mistakes that were boldly proclaimed as truth, including the worst hoax in the history of science.

In this case, as with the others, living animals that nobody believes are our ancestors could be lined up to look like a series of human ancestors. Evolutionists have often chided creationists for using short-hand descriptions or mocking caricatures as "Monkeys aren't our grandparents," or "We didn't come from apes." The actual theory is that living monkeys and apes share a common ancestor with us – however, the living creatures are much like expected stages from more primitive prosimians to humans, and if all we had to illustrate how we evolved were fossils of monkeys and chimpanzees (especially if none were still alive) then they would be put into service as illustrating something like what our ancestors must have looked like.

One of the first proposed ape-like links between us and our ancestors was Neanderthal man. The presentation of these beings as half-witted knuckle-draggers continues to haunt popular thought, although much evidence has since shown them to be very much like us, if not the ancestors of anyone now alive. Another fossil type that was sometimes presented as a primitive "caveman," Cro-Magnon Man, has been known for a long time to be anatomically modern humans living in primitive conditions.

By the time of the Scopes trial, one of the stars of the evolutionary story was Piltdown man, and a leading scientist strongly mocked William Jennings Bryan for representing the creationists even though another "missing link," Nebraska Man, had been found in his own state. Ironically, Nebraska Man, which never consisted of anything more than a single tooth, turned out to be a sort of wild pig. Later still, Piltdown man was shown to be a hoax. No other outright hoax has ever been so strongly and widely accepted by mainstream scientists.

This left just two fossil types represented by fairly extensive remains that could be argued to have served as links between an extinct ape (or very ape-like animal) and modern humans: australopithecines and *Homo erectus*.

Parenthetically, it should be noted there are fossil "human ancestors" assigned dates older than that of the australopithecines, but they all fall into two categories that make them less worthy of consideration. The first and most common is the group of fossils of various extinct monkeys and apes that are obviously very far from being human. With any such group, some will appear somewhat closer to humans in some way or another. The other group is fossils which show some surprising "advance" (greater similarity toward the human condition) in one or two traits, but of which only one or a very few partial, fragmentary specimens have been found. While interesting, for all we know these animals were highly specialized or very "primitive" in other parts of their bodies, and not at all close to the supposed evolutionary line toward humans. It's even possible they were unique freaks or part of small populations that went extinct without leading to anything else. There are also a few specimens that are considered somewhere near the line immediately preceding *Homo erectus*, but these, too, are few and generally scanty and fragmentary.

So *Australopithecus* is the only fossil kind of which we have a good amount of fossil material and appears to be a step beyond any other ape in being similar to humans. There seem to be two main claims for this similarity. One is the structures of the hips and knees that indicate an upright stance, and the other is the apparent use of crudely chipped stone "tools."

While the ability to walk upright habitually is unique (as far as mammals go) to humans, a fossil ape with this ability would not necessarily prove a link between apes and humans. The fossil evidence linking australopithecines to other apes is fragmentary, and likewise there are significant differences between them and *Homo erectus*. It may surprise a lot of people just how large that difference is and how likely it may be that australopithecines couldn't walk upright very long or very well. I say this because popular presentations of these apes have shown very human-looking (if hairy) beings walking with human postures and leaving straight lines of very human-looking footprints. However, further research has shown that they were not that much like humans in appearance or their bipedal locomotion ability.

One of the more visually striking re-evaluations is the structure of the rib cage. The first reconstructions showed it in the tubular or barrel-shaped form that humans have. More recent studies have shown that it was more like the narrow-topped, broad-bottomed shape that chimpanzees have. This means that their arms could not hang straight down and swing straight forward and back. In humans, this action counters the forces of swinging legs to provide efficient, balanced motion. Lacking this ability, australopithecines would have had a waddling gait, more like that of chimps.

The evidence that they could walk upright for any significant distance is rather slim and uncertain. It all depends on the reconstruction of one or two specimens of the hip and the knee, and these are somewhat subjective and subject to controversy. At least one australopithecine skull was examined by X-rays and the space for the organs of balance (the semi-circular canals in the inner ears) was oriented the same way as in other mammals that walk on four legs. Features of the fingers, toes, and wrists also show signs of adaptation for moving by swinging on branches or using the front paws (knuckles) for walking. Combined with the small brains and other features of australopithecines, it seems these creatures weren't nearly as human-like as most popular depictions have suggested.

But what about the tracks showing very human-like strides and footprints? Well, they didn't find the fossil of an Australopithecus at the end of them. They are so human that if they had been found in rocks acknowledged to be recent in formation, nobody would doubt that they were made by humans. The only reason they are ascribed to australopithecines is that the rock was assigned an age much older than that given to any human fossils. If the rocks had been found in, say, Mexico, they would have been re-tested to check the date. But with the hope that australopithecines could walk almost as well as humans, they don't challenge the dating or propose that humans evolved at such an early date.

Likewise, the evidence for making tools is based on the assumption that the "oldest" ones are too old to have been made by humans. These tools are small, chipped stones that arguably show little advance over the "tools" chimpanzees make by stripping the leaves from twigs or branches; on the other hand, it's arguable that they are a major advance in that they require using another rock to make them, and a considerable degree of visualization, manual dexterity and coordination to make the chips just right.

There are a number of reasons to doubt that australopithecines made these tools. Beside the YEC arguments that the dating techniques are completely misleading, there are a number of cases in which evolutionists have re-dated findings by significant percentages. Also, it's logical and generally believed (in the evolutionary view) that species evolved before the oldest fossils. So it is usually assumed that a species appeared much earlier than the oldest fossils. This suggests it is possible that humans were around at the time and made the tools. Also, the small brains and more ape-like hands of australopithecines argue against their ability to make even simple stone tools. Those who have re-created the tools have found that it takes some care in selecting the rocks, and a good deal of coordination and practice to chip them properly.

So unless we find an Australopithecus fossil with a half-chipped rock in one hand and another rock in the other, the association will always be uncertain. Consider the case of Zinjanthropus, which was once called "Nutcracker Man" because of its large jaws and the belief that it had created the stone tools found near it. This made it seem like a possible ancestor of man. Further studies, however, discredited the notion that Zinjanthropus could have made the tools, and it now has a different name and is placed on a side branch of the australopithecines, a dead end that evolutionists agree is clearly separate from human ancestry.

Another curious aspect of this scenario is that very similar tools are attributed to early fossils assigned to our genus. Is it credible that so much time passed and so much evolving occurred, and yet no advances were made in the tools? And what about these "early" hominid species? Do they show close affinity to the ape-like australopithecines?

The "early species" of genus Homo (such as *Homo habilis*) are represented by a few fossils which are not very impressive, considering how little of the skeleton is represented and the poor condition of the bones that are available. If you divide the available fossils into different species such as *Homo rudolphensis*, there is very little material per species. But right on their heels comes *Homo erectus*, with many widespread cases, which in many aspects show the appearance of modern human traits.

While australopithecines have been shown in artist's conceptions as standing straight upright and looking almost like small, hairy humans, *Homo erectus* (if shown at all) is usually shown as just a fossil skull. A side-by-side comparison of accurate reconstructions of their skeletons, along with modern chimpanzee and human skeletons, shows a great leap of difference. (See Nature, 2004 Nov 18;432(7015):345-52, "Endurance running and the evolution of Homo," Bramble DM, Lieberman DE. Trace or copy onto clear plastic the skeletons illustrated on p. 349, cut them out and superimpose them for the full effect.) The australopithecines were very chimp-like in overall size and shape, while *Homo erectus* was very much like a modern human. Unlike the australopithecines, *Homo erectus* was built (not just for walking but) for long-distance running, as we are, although relatively few of us live up to that potential. The only significantly "primitive" aspect is in the thick, beetle-browed skull with a smaller brain case. But does this mean that they were less intelligent than modern humans?

It would not be inconsistent with the creation framework if there were a very human-like animal, or a line of humans that had mutations rendering them more ape-like and less intelligent. The Bible's book of Job includes a reference to humans driven from society and living in the wild, and the *Epic of Gilgamesh* describes how a wild, hairy man, Enkidu, was drawn away from the wilderness to live in the city. On the other hand, there is also the possibility that the different skull of *Homo erectus* housed a brain on a par with our own. When it comes to brains, size may be a clue, but it isn't everything.

Unlike the australopithecines and the rarer remains assigned to our genus, *Homo erectus* fossils have been found in southeastern and even northeastern Asia as well as Africa. Dates assigned to their fossils overlap those given to Neanderthals. This indicates they were very successful at traveling, spreading out and surviving in different environments: hot and cold, dry plains and humid jungles. At some sites only primitive stone tools such as those attributed elsewhere to australopithecines have been found, but at other sites, much more advanced artifacts have been found, including traces of huts and controlled fire. The sites in northern China would have required such things due to the cold, and one or two sites on islands indicate the ability to navigate stretches of ocean.

Of course, there's no telling how much of the data or "facts" about *Homo erectus* (and fossils with far less known) is due to misinterpretation or outright fraud. We'd like to think that scientists are not susceptible to such things, but they are still human, after all. There are known cases in science (especially this area) of mistakes, fraud, plagiarism, and other unethical practices. And for all we think we know, there is much we know we can't know for sure. So it may be that *Homo erectus* was just an unrelated if remarkably human-like ape, or a tribe of *Homo sapiens* that degenerated and died out. Or maybe they were the ideal humans, far more intelligent than us but struggling to survive after a global catastrophe, and we're their children with oddly mutated, enlarged, rounded skulls. Evolutionists and creationists alike can argue among themselves as well as against each other about the exact status of such cases, but it should be clear to all who are willing to doubt the claims of evolutionism that there is a distinct gap between australopithecines and our genus, that is "filled" only with bits and pieces and a lot of questions, not a clear lineage between apes and humans.

Point Summary: Plausible fantasies are bad science, not facts.

It's understandable that evolutionists are emboldened by these cases. In a general way, they do fit the evolutionary framework better than the creation/Flood scenario. However, none of these cases show a gradual development from one population to another. Instead, they show (at best) a few samples from specialized side branches of the expected lines from a given ancestor to modern forms. Only a few exceptions are regarded as probably in the actual line of descent. The apparent consistency between theory and data depends on accepting a number of excuses for why the data doesn't clearly match the theory.

If evolution is a natural, gradual, continuous process, we might expect a sampling of fossils that depends on natural forces rather than intelligent or systematic selection to present different stages of evolution of all organisms present at nearly every site, requiring many new classifications at each one. Instead we find recognizably distinct kinds, including extremely specialized forms. While the excitement and prestige of finding something new has led to the naming of many species of fossils, there aren't very many distinct forms of animals and plants. Some fossils dated hundreds of millions of years old are almost identical to living animals and plants, despite all the changes in the environment since they were formed.

Another inconsistency is that some geological formations appear to show the sudden appearance of highly specialized animals or even a wide range of new forms. Sometimes these correlate roughly with some global change, but sometimes it is reckoned that the rapid evolution of new forms began significantly before or after a change. In other words, the few samples of fossils that seem to roughly fit an evolutionary development are more than matched by cases in which very distinct forms show a great deal of difference from the proposed intermediate or ancestral type. In other areas of science an assertion (hypothesis or theory) must first be demonstrated to be possible before it gains much respect. Then, if even one case doesn't fit the assertion, it may be discarded or at least re-evaluated and subjected to further observations and tests. In evolution, nothing more was required than leaps of imagination, from the observation of slight variations in living animals, to connect all of life through an admittedly too-small sample of fossils. After that, every new more-or-less intermediate fossil has been proclaimed as proof of an evolutionary transition. It's like a game of connect-the-dots, where you only have dots and assume they are connected to form a tree – but just what the tree looks like might be greatly revised with each new dot discovered. Meanwhile, cases where transitional forms are strikingly absent continue to be downplayed merely through the hope that eventually transitional forms will turn up. But let's look at these cases and consider, to the contrary, whether they may indicate that the cases of supposed transitional forms are the result of wishful thinking, for the most part.

Next point: Hey, These Dots Are Really Far Apart!

Starting again with the case that is the "oldest" by conventional dating (without going back to the question of the origin of life itself), we find that the most startling case of lack of evidence for evolution is that of the appearance of all the major divisions of life! There are a few traces attributed to the presence of life dated so far back (approaching 4 billion years ago) that the Earth is thought to have been barely habitable by any form of life. These are mostly lumps formed of layers of tiny grains that look like the piles of sediments trapped by bacterial colonies which form today, called stromatolites. Rocks given dates of lesser age, up to about 500 million years, show only rare traces of the same sort plus some mysterious, controversial tracks and other tiny traces that don't reveal much. There are completely different traces known as the Ediacaran fauna, and tiny bits of shells, dated a little earlier. But in the rocks assigned to the Cambrian period, many different kinds of life are found. They are all bottom-dwellers, bottom-feeders and weak swimmers, but they represent all of the major phyla of animal life. There are also a number of fossils that represent totally different, extinct phyla. Of course, evolutionists see in the Ediacaran forms some indications of ancestry leading to Cambrian life, but they're pretty much forced to do so, not having anything else of significance to work with for over 150 years. This vast diversity appearing with so little coming before it is so startling it is called the Cambrian Explosion. Evolutionists have come up with stories as to why ancestral forms didn't leave traces and why life would have so rapidly evolved in so many very different ways, but these are little more than lame excuses which show the psychological need for faith in evolutionism, or the philosophical straitjacket that has been put on this extension of science.

Darwin's view, and that of most evolutionists ever since, was that evolution was a constant, gradual process of nature. Some evolutionists eventually began to consider that the fossil record wasn't merely missing some parts, but that evolution proceeded in fits and starts, sometimes happening so rapidly that it was unlikely for the transitional forms to be preserved. However much popularity this view does or doesn't currently enjoy, the fact remains that different rock formations show strikingly different sets of fossils, and evolutionists generally hold that there was some degree of "rapid radiation" involved. One aspect of this that is downplayed (to the point of being removed from presentations to the public) is that this sudden appearance occurs in cases shown to the public as gradual transitions, as noted for the dino-bird and reptile-mammal cases.

Although a few Cambrian fossils look like simple fish (no fins or other such details appear), there are not many fish fossils in rocks from there through several other divisions of the geologic record. In the Devonian rocks, however, there is an abundance of fish fossils, and again the major differences are represented: Fish with cartilaginous skeletons like sharks, fish with fins on the ends of flesh-and-bone paddles like coelacanths, and fish like most modern fish, with thin fins attached directly to their bodies.

There's no sign of a gradual evolution toward amphibians, only a few supposedly intermediate forms, a gap with few amphibian forms, and then suddenly a wide variety of amphibian fossils, including some very specialized and large forms. All of them are extinct, with modern amphibian fossils showing up much later, and again with only imaginary connections to the extinct forms.

There are a few fossils supposedly showing some stage or other in the evolution of amphibians to reptiles, and then again a wide variety of definitely reptilian fossils show up. Some of the earliest are those called mammal-like reptiles. In the first division of the Mesozoic (the Triassic), dinosaurs show up, and so does the first fossil identified as a mammal. The earliest dinosaurs include both bipedal carnivores and quadrupedal herbivores, but it's not until the Jurassic that dinosaur and mammal fossils show a new level of diversification.

While the Jurassic and Cretaceous mammals are all fairly small (no larger than a badger or a bulldog), their diversity includes everything from a semi-aquatic form to one that could glide like a flying squirrel. However, few if any of these mammals are supposed to be the ancestors of modern forms. Only some little creatures without specializations are supposed to be closely related to the ancestors of modern placental mammals. Fossils of modern forms only appear in Cenozoic rocks. These fossils have some differences from living mammals, but they are clearly the same types. Such distinctive differences couldn't have evolved overnight, and so it is believed all the forty or so modern orders must have actually evolved in the Mesozoic, but left hardly any fossils that we've found in those rocks.

Likewise, in spite of the confident claims of birds being living dinosaurs, none of the birds of the Mesozoic which are supposedly evolved from dinosaurs appears to be directly ancestral to modern birds, but the great variety of modern-appearing birds in fairly "early" Cenozoic rocks leads to the conclusion that the forty or so lineages of modern birds had their origins in the Cretaceous.

The sudden appearance of many different flowering plants (angiosperms) in the Cretaceous is another apparent rapid radiation. There are probably a number of lesser cases that could be cited, but have been easily overlooked or ignored by researchers, or simply not presented to the public. Within the mammal orders, for example, the elephant type is represented by a series that is mostly imagination connecting the dots between early "primitive" forms (which might well represent distinct types with no relation to living elephants) and essentially modern forms like mammoths and mastodons, with "side branches" represented only by the highly specialized varieties presumed to be the tips, such as the shovel-tusked elephants. Then there's the apes, which also show up as distinct types recognizable as slightly different from living forms, not a collection of puzzling intermediates, so that evolutionists again have to connect an array of dots with imaginary lines. There's a particularly striking lack of immediate fossil ancestors of orangutans, chimpanzees, and gorillas – have some of their ancestral fossils been assigned to the much more exciting and prestigious line leading to humans?

In addition to these cases of broad diversity showing up with few if any fossils of actual ancestral types, there are several specific kinds of animals that are highly specialized, and there are no good fossils showing any transitional stage towards these extreme forms. One of these is the turtles, with their very distinctive shells. While some turtle fossils appear to have a "primitive" feature or two, they all have fully-formed shells. There are some reptiles with somewhat flattened ribs, and one or two badly-preserved, fragmentary fossils appear to have only half a shell, but that's not much to go on for such a distinctive condition with many forms involving a hard shell rather than just soft tissues. No other reptiles have extremely large scales or partial shells that represent intermediate stages. There's nothing close to a series (or rather, several series) of intermediate forms showing how all the different tortoises, freshwater turtles, and huge sea turtles evolved. It doesn't help at all that there are fossils of reptiles which were somewhat turtle-like in their flattened forms covered by thick pebbly scales, because they are clearly not ancestral to turtles and only show that if there were any ancestral turtles, there's no reason why they shouldn't have been preserved as fossils and found also. They further suggest that the few posited intermediate forms might also just be unrelated reptiles with some similarities.

Flight is a special problem for evolutionists. While there are feathered dinosaurs they can believe are what modern birds evolved from, none of the other animals with the ability to fly have any such possible transitional forms. Flying insect fossils are dated to the time when fish were supposedly evolving into amphibians. There are fossils of huge dragonflies dated around 300 million years old. And insects have several distinct forms of flight. Consider how different dragonflies, butterflies, and beetles are. While insects have the advantage of being lightweight, they also have tiny muscles and brains, and being so small means the air is like syrup to them. It takes special designs for them to fly at all, yet dragonflies are capable of flying in all directions and hovering, and can fly very fast. Evolutionists have come up with clever stories about how flightless insect ancestors might have evolved flight, but they can only go by observations of living insects that can fly as adults.

Pterosaurs are highly specialized for flying, with their wings largely supported by one series of long bones comparable to our pinkie fingers in their relationship to the other digits. There are no fossils of anything else that has this digit significantly enlarged. All of the pterosaurs have this feature (and others, such as the wing membrane) developed so far as to have made them perfectly capable of flight. None of the handful of proposed ancestral forms could so much as glide using membranes attached to their forelegs. Some evolutionists have been so desperate as to suggest the ancestors are represented by fossils of small lizards with membranes attached to long hind legs! There are other fossils of lizards that could glide in various ways, but these are clearly not related to pterosaurs and only raise the question of why they were preserved and discovered, but not the imagined ancestors of pterosaurs?

Likewise, bats fly on wings supported by several extremely lengthened digits. Not only are there no fossils of bat ancestors with an intermediate condition, there are no fossils of any animal which had only partly elongated digits supporting a gliding membrane. There's no indication that would be a viable evolutionary path at all. While some bat fossils are said to be primitive in some feature or two such as the length of the tail or lack of echolocation, all of them have fully-formed wings. Not all living bats use echolocation, and one of the "earliest" types of fossil bats was preserved with indications it did possess this amazing ability.

We might also ask why we don't see all sorts of living intermediate forms. One response is that everything is already fully evolved, but according to the theory of evolution, living things are constantly being subjected to changing natural conditions, with new natural selection pressures. Arguably, everything might be in a transitory state and evolve into something very different in the future, but all living things seem to fall easily into a relatively few distinct forms that have been around for millions or tens of millions of years at least. There are a lot of things that seem strange if Darwin's general concept of slow changes gradually accumulating in living things were true. Darwin and evolutionists since have given explanations they found plausible, but when you put all the questions together they defy the explanations.

Why aren't there always many transitional forms blurring these distinctions? Why would whales and other mammals (and extinct reptiles, too) evolve into aquatic forms, but not one kind of fish ever evolved to live on land after the first amphibians? Why are there such distinctions between mammals adapted to various degrees of aquatic life, and why do "transitional" fossils tend to parallel these divisions? Why have no reptiles also evolved into a series of increasingly aquatic forms again? Why don't we see anything like a flying squirrel on the verge of powered flight, like the supposed ancestors of bats that took to the air when birds were already there, or the supposed ancestors of birds that would have started flying in skies occupied by pterosaurs? Why are there no mammals with moderately-evolved bipedal walking like our supposed ancestors. Why have no reptiles again evolved the efficient upright stance of dinosaurs? Why did none of the large "terror birds" found in rocks above the Mesozoic-Cenozoic boundary not evolve into forms more like the missing dinosaurs they supposedly evolved from?

It's easy enough to dismiss such thoughts with the explanation that evolution is random and we can't expect it to fit our expectations, but that suggests that the theory is useless, except as a box for our minds to tell stories about the past, shutting out any doubts or other possibilities. It merely replaces "God did it" with "Evolution did it."

Point Summary: Which is the stronger evidence, the dots connected by imagination, or the glaring gaps?

How willing should anyone be to accept explanations (excuses?) for why series of transitional forms are missing? How long can people have faith that they will be found some day, and still consider their faith a part of science? Wouldn't we expect tricky transitions such as the development of flight to require a long series of random mutational "experiments" with matching environmental changes? Why does practically every case in which some transitional forms are claimed to exist depend on appeals to their representing side-branches that were only something like the true transitional forms, or late remnants of earlier "ghost" lineages? To someone who wants to believe in the power of evolution, perhaps some intermediate forms are enough to support their faith that all of the required evolutionary transitions actually happened, but to anyone with a healthy scientific skepticism, the fossils and the pattern they form do not show this picture, even in a fragmentary way.

Section III. Facts that evolutionary faith and fantasies are up against.

Subsection A: Facts indicating that natural processes cannot produce all the extremely complex forms of life from the less complex forms.

While evolutionists have faith that great mysteries such as the origin of life will be found to have perfectly natural explanations, the facts as we know them now are contrary to that faith. That many highly intelligent people believe in evolutionism in spite of these facts is a tribute to the power of the human wish for all of reality to be as normal, comprehensible, and predictable as our current ordinary experiences.

Let's start with the general evolutionary faith that there exists a biological-environmental mechanism capable of generating all the forms of life from the simplest possible form, given enough time or generations. It's encouraging for scientists to think that ordinary processes comparable to those that produced different breeds of dogs are all that is needed to explain the existence of all the various forms of living things. The facts that have been established, however, contradict this theory. The changes that have been observed over many generations are all neutral or degenerative when compared to the need to produce systematically organized, dynamically complex new parts. The genetic hereditary system and natural selection are conservative. At best they allow for the alternating or cycling of encoded variations of equal complexity, and rarely for restoration of lost features. There is no evidence for an innovative mechanism or process that can produce, over any number of generations, such complexity that didn't exist in an ancestor.

It's not only that no change has been observed that increased the complexity of an organism beyond that of any of its ancestors, but in addition, the tendency to degeneration is so pervasive that it is amazing that any living things have managed to keep from going extinct. Shifts in populations (as in the case of peppered moths) produce nothing new, and if continued to extremes would only result in the elimination of variations within species, or "new species" that are nothing more than limited, isolated subsets of an earlier, highly-variable species.

Even beneficial mutations generally involve some form of degeneration. Bacteria that have become resistant to antibiotics have mutations which (for example) damage the control system of excretion mechanisms. This allows them to kick out antibiotics as fast as they get in, but the uncontrolled or overdriven flushing weakens the bacteria and make them unable to compete with normal bacteria in normal environments, and the change is not a step toward producing a new mechanism, but only damage to an existing system. Although it happens to be beneficial in a specific environment, it is a change in the wrong direction when we're looking for examples of the power of evolution to produce all living things from microbes.

Sidebar: What's all this about complexity?
To bolster their vague concept of mutations somehow producing whole series of viable, significant increases in complexity, evolutionists generally remain satisfied with a fuzzy concept of complexity, and they resist attempts by the Intelligent Design community to clarify it, indicating that evolutionists sense it will prove to be a more considerable hurdle than they care to face. Instead, they seek to find a general natural

principle of "self-organization" that depends heavily on examples of simple ordering such as the formation of crystals or the pattern of swirls in boiling water.

While "complex" is applied commonly to many different things, and changes from one context to another, it should be possible to develop a clear and useful concept of physical complexity. More complex things have more parts than less complex ones. The parts are made of different materials. The parts have different shapes. The parts can move. The parts move systematically, not chaotically (they move only in certain ways, limited by other parts). The most complex things systematically harness energy to move on their own, and can actively and systematically transform energy and/or materials. Living things sit on the top of the complexity pyramid with the additional features of actively and autonomously gathering energy and materials, growing and repairing themselves. They can also copy the data that describes their construction, split off parts of themselves with this duplicated information inside, and thus reproduce themselves.

A rock is a common example of something simple. It may technically be composed of many different minerals, but they are amalgamated into a single unit that doesn't do anything. Drilling a hole or otherwise shaping the rock may give it a somewhat more "complex" shape, but (with a special class of exception noted next) no matter how many holes you drill or how wildly you sculpt its borders, it will still be a simple object.

If you break a rock, you will have two simple objects, but as separate objects there is no increase in complexity. However, an intelligent and very skilled sculptor with the right tool might be able to carve the rock in an intricate way so that it was in two pieces still joined together, like pliers or scissors. This might be considered a basic unit of complexity – two parts which are effectively joined as a unit but able to move (or rather, to be moved) in a systematic way.

So we note that complexity can be recognized and different levels of complexity can be measured, but not by any simple means such as counting things or comparing something to a yardstick or other device. Different materials or shapes may not increase complexity significantly but they can contribute to it. Although more parts are required for more complexity, adding more of the same part does not increase complexity significantly. How an object is used or how many things it is capable of being used for may not say much about its complexity. With a bit of imagination, simple things can be said to exhibit advanced traits of complexity. However, if we avoid stretching definitions, there are some traits that apply only to the most complex things.

One aspect of complexity has been given the name "irreducible complexity" by Dr. Michael Behe. As with the general argument for recognizing intelligent design, evolutionists often give this the over-simplified description of "saying that life is too complex to have evolved." Ever since Darwin, evolutionists have depended on assuming that all change, and all complexity, can be described in terms of simple degrees: as if "less complex" and "more complex" were terms equivalent to "lighter" and "darker." When properly understood, irreducible complexity emphasizes that it's not merely the number of parts or the intricacy of their shapes or arrangement, but the need for a minimum number of parts organized to work together in a certain way to achieve the function of the whole. Behe's classical illustration is a mousetrap.

The evolutionists' responses, such as the creation of trap-like devices with fewer parts, or using a broken mousetrap as a necktie clip, were produced by human intelligence and do not show that random changes could produce such a series. (And many cases of the simplified mousetraps require faith that they would do anything useful, or that ancient mice would be very easy to catch.) If such a series could be produced by random changes, the various stages would all have to have viable functions. In actual biological cases, evolutionists assume that cellular or larger organic mechanisms with similar but fewer parts represent or illustrate intermediate stages toward the more complex forms, but they haven't shown that mutations can produce any such series, let alone a series with some naturally-selectable advantage at each step. In other words, evolutionists again have great faith in their theory and so are satisfied with vague, wishful imaginations of what might have happened, in spite of the fact that no such thing has been shown to be possible.

Another aspect or description of the kind of complexity found in different parts of living things is "specified complexity," championed by Dr. William Dembski. This formulation emphasizes the fact that life requires its parts to be arranged in a certain order, even when they're just strings of nucleotides in chromosomes, or amino acids in proteins. It shows that there can be informational complexity in something that is relatively simple physically. Although DNA is essentially just a string of four chemicals, there are far more possible arrangements of them that are non-viable than those that can have a useful function. This makes the odds of random changes producing a more complex system far too low to be believable. (I would note also that it takes a great deal of physical complexity to translate the code into viable proteins without mistakes, identify and destroy faulty proteins that get made despite the quality control features, etc.)

The evolutionary response to the concept of specified complexity is that since evolution works through small changes, there is much greater probability of success at each small step, and so the total change is accomplished by numerous probable steps rather than one or a few impossibly improbable ones. The problem with this is that it assumes there will be random changes to genetic coding that produce an entire series of simple changes that still make a difference in natural selection (providing a significant improvement in the body) and eventually combine in one of the relatively few ways that produce a useful form of greater complexity. To optimistic evolutionists who focus on the existence of beneficial mutations this may seem plausible, but those cases almost always show some loss of complexity. Even if we were to cheat a bit and say that mutations restoring parts that once existed in ancestors are *new* increases in complexity, there would still be huge barriers for such small steps to overcome. Consider the fantastic layers of complexity and the broad diversity in living things, and the extensive, precise genetic coding required to specify all those parts! Then there are all the harmful and even beneficial mutations which are constantly reducing complexity. It takes a special kind of faith to believe that there wouldn't be any insurmountable hitches in the process. Mind-boggling complexity and diversity are facts of life which the theory of evolution cannot explain to the satisfaction of anyone not committed to purely naturalistic explanations.
End sidebar

I like to say that we can be certain an object was intelligently designed if it has systematically organized dynamic complexity – diverse parts connected to interact to produce change in a way that requires the regulated actions of all the parts. This leaves out a lot of designed things, but living things are clearly more complex than this. I believe there is an obvious gap between technological devices which meet this minimum requirement and anything that occurs naturally in the non-living world. No example of "self-organization" in nature comes close to producing such complexity, let alone what's required for life.

Living things have a kind of complexity that is higher than that of any artifact, in that they are self-constructing, self-sustaining, self-repairing, and self-reproducing. Living things are part of nature, but their origin was not observed by humans. The scientific theory of evolution only deals with variations in living populations, but when it comes to evolutionism vs creationism, how life and all its forms arose is of vital importance. Then there's the question of how all the systems in living things supposedly arose from the hypothetical first "simple" living things. Intelligent design is the only observed means of producing the systematically organized dynamic complexity that mere parts and sub-systems of living things display.

We can observe ourselves intelligently producing things with new examples of irreducible complexity, specified complexity, and organized dynamic complexity. We do not see this happening in nature. Living things have greater complexity, but as we observe many generations we see only replication of previously existing complexity, with a trend toward loss, not increase. Living things thus appear to be designed by an intelligence greater than the accumulated knowledge of all humanity. Life is amazing in its unique abilities, not found even in our most advanced technologies.

Subsection B: Facts contrary to the supposed appearance of common ancestry

One of the main arguments for evolution has been that we can see similarities and differences in our own families and in varieties of domesticated plants and animals, cases in which we can trace different groups back to known common ancestors. Darwin was especially impressed by the differences in breeds of pigeons. There also seems to be a broad range of variation in breeds of dogs and of horses. By the logic of analogy and extrapolation, it makes sense to attribute the similarities in all living things to their all having a common ancestor, too.

However, when we look closer, we see it goes beyond reasonable extrapolation and is more like jumping to conclusions – jumping from changes in degree, size or shape and many close similarities to many changes requiring new complex parts and great differences. Another problem is that there is no real way to test its validity. It is said to require far more time than the entire history of civilization for new species to arise and become established, and speciation is very different from the formation of the differences between higher levels of classification.

Furthermore, this idea that similarities are generally due to common descent, when combined with the belief that evolution can produce any given difference between living things, can vaguely explain any pattern of similarities and differences with apparent ease, especially considering the similarities required for any form of life to survive in harmony with others in the same environment. When no other explanation is allowed, some variation of this is inevitably going to be accepted, at least on a working basis. If things that are supposed to be closely related are discovered to have surprising differences, evolution can be said to have operated faster than usual. If things considered very distantly related have genes or other parts that are amazingly similar, they can be said to have been molded by similar environments or to have retained the similarity from a common ancestor while all other related forms lost it. If the explanations get too complicated or extreme for comfort, the branches of the Tree of Life can be re-arranged until the stories required seem more plausible.

Believing that somehow life itself arose by natural processes, evolutionists can always shrug and say something just happened to happen, just by sheer random chance. Thus, any case that might be proposed to test the theory will always be explained away one way or another, without ever demonstrating that the explanation did, or even could have, happened. This would never be acceptable in the practical areas of science that we all admire so much.

The supposed appearance of common descent (similarities), is further weakened by the fact that created objects also have varying degrees of similarities, even when designed by different people and for different purposes. Artists produce series of works with similarities based on a theme or motif. Cars made by the same manufacturer have similarities in style and trivial, non-essential details. Cars, boats, and aircraft have similarities in their motors, while tricycles, bicycles, motorcycles, cars and aircraft have similar wheels. It's possible to create cladograms or evolutionary trees with all sorts of created objects. Why should it be any different for living things, if they are also created? Only a madman might create things so randomly that they couldn't be classified in an orderly way.

There are many cases about which evolutionists themselves say that the similarities are due to "parallel" or "convergent" lineages rather than retention of traits from a common ancestor. If evolutionists recognize the fact that a number of similarities are due to design requirements (such as environmental constraints on survival) rather than common ancestry, how can we say with any confidence that any similarities are due to common ancestry, beyond the sorts of variations that have been directly observed?

There are many common (and arguably trivial) cases of independent similarities in overall body shapes and colors, such as in various green tree snakes, various predators that live in grasslands, the various prey species, etc. A number of animals have a crab-like appearance, from the various crab spiders which are named for their vaguely crab-like appearance, to creatures such as the porcelain crabs, that only experts can distinguish from true crabs.

More intriguing is the similarity of parts of animals that are otherwise quite different. The robber crab, which can't live in water, has a system of smelling which is similar to that of insects. The similarity between the heads of insects and centipedes isn't attributed to common descent. The eardrums of frogs, of birds, and of mammals, must have separate origins.

Some features seem designed to defy the notion of shared ancestry. The different molecules that give blood its pigment and oxygen-carrying ability present a patchwork pattern. The hemoglobin of our red blood also appears in common earthworms, snails, sea cucumbers, some mollusks and insects, and other lower animals not imagined to be close to us on the evolutionary tree. Haemocyanin, which can make blood blue or green, is found in some arthropods (lobsters, crabs, crayfish, and scorpions) but also in some mollusks (clams, mussels, octopuses and squids). Some worms that live in the sea also have hemocyanin, but the violet-colored Hemerythrin is used by other worm-like sea creatures and the clam-like brachiopods. Chlorocruorin, which uses iron and is red like hemoglobin, is found in other worm-like sea creatures and also in some starfish. Some kinds of fish have molecules in their blood that act as antifreeze. One particular kind of antifreeze molecule is found in different fish at opposite ends of the Earth, the arctic and antarctic.

There are also cases of "convergent" features that link fossil and living forms not believed to be closely related. A number of fossil reptiles, such as the champsosaurs and Goronyosaurus, show a great similarity to crocodiles and alligators. Some dinosaurs appear to have had beaks designed to filter tiny prey from muddy water the way duck beaks do. Flamingos have similar beaks, except they are designed to be used upside-down. The pterosaur Pterodaustro also had a beak designed for filter feeding, but it was extremely long and filled with many long, thin teeth. Another kind of filter-feeding is found in aquatic animals from ancient jawless fish to paddlefish, whale sharks, manta rays, and more. Baleen whales have their own, highly specialized, form of filter feeding.

Fossils have indicated that teeth would have had to evolve more than once, and the extremely long canine teeth of the so-called "saber-toothed tiger" also appeared on some true cats, hyena-like animals, and a large cat-like marsupial (all of which are now extinct). Marsupials present many other cases of convergence. The wombat is rather like a badger, and there are (or have been) marsupials very much like mice, moles, wolves, small cats, and lions. And, like primates, birds, and some other animals, some marsupials also have full (three-cone-type) color vision.

Similarities in birds provide some surprises. Sunbirds appear and behave much like hummingbirds, but further analysis suggests they are closer to swallows, while the swifts which seem so much like swallows are closer to hummingbirds. Vultures in North and South America appear most similar to storks, but eastern hemisphere vultures are more like eagles. Owls are predatory birds like falcons, but their biological and behavioral differences are such that the similarities don't appear to be due to common ancestry.

A couple of amazing cases of convergence don't involve detailed similarity, but show the design requirements of certain lifestyles, and raise the question of how random mutations (even guided by natural selection) could produce so many similar traits that meet such extreme and stringent conditions.

The ability to fly is one such case. In the case of insects alone there are two or three distinct types of wings with different flight characteristics. Pterosaurs, birds, and bats each have their own unique wings. Birds also happen to be the only flying vertebrates known to have flightless forms – and they have a number of successful flightless forms. Extinct feathered vertebrates that had lost the ability to fly (or were created flightless) would look like intermediates.

The ability to glide or parachute in a controlled manner is another case. There are a number of fossil lizards (and one living kind) clearly designed for this, several of them with long ribs and several with unique designs. There's even one (and only one) Family of flying fish which can extend their leaps out of the water by gliding with the aid of their extended fins. Some squid are also able to jet into the air and travel some distance. Some tree frogs have feet and bodies which allow them to safely slow and direct their falls after leaps from danger in tall trees. One or two types of tree-climbing snakes can flatten their bodies and control steep glides, also. Then there are different species of flying squirrels, and also flying lemurs (Colugos) which have quite impressive gliding abilities, but are not lemurs nor closely related to them. Significantly, in relation to the belief that flying things evolved from non-flying ancestors, none of these gliding forms have soaring or flying relatives.

The ability to swim efficiently and live in the water for the entire life-cycle is another case, of course, although supposedly the ancestors of all living things were aquatic. General features of streamlining and propulsive fins are seen in a number of lines in which their appearance cannot be attributed to common ancestry. Fish, mollusks, reptiles, mammals, birds, and insects, in a number of distinct forms in each group, show the basic similarities. In some cases the similarity goes a good deal farther. Certain species of tuna, sharks, some other fish, and the extinct reptilian ichthyosaurs show similarities in their very streamlined forms and propulsive tail shape, and so do some whales except for the horizontal orientation of their tail flukes. Some of the sharks and tunas are similar in their internal designs as far as the propulsive musculature and related features are concerned. Whales, dolphins, and seals share techniques and designs for deep diving that allow them to push the envelope of the strict limitations imposed on air-breathing animals.

The ability to live by eating ants and termites almost exclusively is another case where several different mammals show highly specialized shared design features. Giant anteaters, spiny anteaters (which are marsupials), tamanduas, aardvarks, and pangolins all exhibit the long snout, long sticky tongue, and large digging claws, along with digestive system traits, that are needed to gather and digest sufficient nutrition in this manner. In this case, too, there is little, if any, evidence for gradual evolution of the distinctive traits. On the contrary, the few fossil forms that have been found look very much like the living ones.

Remember, the fossil record shows almost all the living orders of mammals – all the major differences in design – appearing independently, with little or no sign of the imagined dozens of transitional lines (and vast numbers of transitional stages in each line) that would have existed if they descended from a common ancestor such as one of the Mesozoic mammals. More surprising is the similarity in the organization of mammals' and birds' brains. Songbirds share with humans a special molecule involved in brain development (it is required for vocalization). Octopuses have eyes that are similar to those of mammals in many aspects other than the arrangement of the retina. For that matter, "The biological clock of the honey bee is more similar to that in humans than to the one in flies" (and other insects - Press release from The Hebrew

University of Jerusalem, 25 October 2006, regarding an article appearing in Nature 10-26-2006).

Then there's the case of animals that generate electrical fields or light. Neither of these are required for any living conditions (though they can be very helpful), and they both require unusual specializations. Quite a few different fish, marine and freshwater, generate electrical fields that they use to sense their environment, and in some cases generate currents powerful enough to stun prey or enemies. Animals from bacteria to fireflies generate light with high efficiency using special chemicals which humans have only relatively recently learned to use (e.g., glowsticks) – by studying the biological examples.

One argument against viewing similarities as due to common design needs (and themes) was the similarities found in "junk DNA" and genes that share the same disabling mutation. It doesn't make sense that God would create different forms of life with the same useless parts (or coding), and (it is argued) it doesn't seem likely that the similar parts would break down in the same way. However, we can see that even the best designs of complex machines may have similar parts which are more likely to break than others. The engines and wheels of cars of different makes, and even of different kinds of vehicles, will break down in similar ways. Also, some similar parts are less important than others and so can similarly break down while allowing the machine (or living thing) to remain sufficiently functional. So, too, similar "vestigial organs" in different kinds of creatures may have degenerated from separately-created functioning forms. Likewise, recent and ongoing studies show that chromosome segments are not equal in their tendency to mutate, so it is not so unlikely that similar coding sections could have similarly mutation-prone sections and so be likely to become inactivated independently.

Furthermore, we now know that sections of DNA once considered to be "junk" do have functions, and more are being discovered, so similarities in others may also be due to functional design rather than common ancestry. These studies also may dampen enthusiasm for the "molecular clock" theory. In the application of this theory, differences in supposed junk DNA supposedly inherited from a common ancestor have been used to "establish" the time when two populations diverged from each other. If they have not been degenerating gradually and randomly from the same familiar protein-coding genes, then their differences can't be used to estimate time.

When looking at all these facts – many cases of similarities that nobody believes are due to retention from a common ancestor, the trend of discovering useful design in "junk" DNA and "vestigial" organs, the fact that good designs can share weaknesses – the faith that similarities can be traced back to ancient common ancestors is seen to be an arbitrary materialistic alternative to the faith that a Creator provided similar good designs to various kinds of living things, which had similar unavoidable weaknesses and variability. The impression or opinion that the evolutionary view is superior again seems to be a matter of faith, instilled by personal preference or indoctrination.

Similarities that can't be chalked up to common ancestry clearly undermine the supposed evidence for evolutionism. It may then seem strange that the great number of similarities that *are* assumed to be due to common ancestry can demand a lot of faith. How can this be? Because these many similarities appear in living things that are very different, clearly distinct, not closely related. For so many living things with such differences to have so many similarities due to common descent would mean that an early ancestor had all these traits, and they remained through long lines of descent. That means a lot of traits would have to have evolved fast, early on, and survived the forces of natural selection through major extinction events and periods of rapid change.

Recall our review of the story that evolutionists believe connects the dots of fossil data. Billions of years with many cases where sets of widely different fossils not seen in lower strata show marked contrast with the belief that they gradually diverged from a common ancestral population. There are also five points in the fossil record above which a great number of forms are never seen again, such as the lack of all dinosaurs, pterosaurs, ammonites and others in Cenozoic strata.

Now, in addition to there being many forms alive today that appear very similar to forms dated tens or hundreds of millions of years ago (and billions in the case of stromatolites), that certain similarities are cited as evidence for common descent requires belief that they all also evolved up to billions of years ago, and have remained the same as the organisms continued to diverge through all that time and all those major changes in the environment. Evolutionists have been surprised at the number of cellular traits that must have been present in the very first form of life, according to this doctrine. Our own bodies would be filled with living fossil parts.

Subsection C: Facts contrary to faith in the natural origin of life

While the origin of life is not part of the scientific theory of evolution, if the naturalistic philosophy that governs current science is consistently applied, there is no essential dividing line between life and non-life. There are also a number of evolutionists who do believe that the term "evolution" covers everything all the way back to the formation of the first stars and galaxies. If it doesn't, there's no sense in many of the grandiose claims they have made for the importance of Evolution.

However, the facts of nature show that evolution couldn't get started. There's the fact that no one has ever observed chemicals spontaneously evolving into a living thing, the fact that nobody has even observed chemicals naturally forming a systematically organized dynamic system, the fact that complex organic molecules are far less durable than simpler ones, and the fact that changes caused by natural, non-directed application of energy are degenerative to anything less complex than advanced technology or a fully-functioning life form. These and other facts present a solid barrier to the origin of life except as the creation of a highly intelligent agent.

Why do so many scientists believe in something contrary to so many facts? In short, they have no choice, because the current definition of science requires that totally natural explanations must be provided for everything that ever occurred in the universe. Almost all evolutionists attack any suggestion that anything other than natural forces affected the universe at any point in time or space. They will not allow consideration that something more might have been involved, and they crusade against questioning or skeptically examining evolutionism. It's a pity that this monopoly of materialistic philosophy has masqueraded as science so long.

Consider some of the details of this case of the origin of life. First there's the general observation that complex things almost universally become less complex over time when exposed to natural energy sources. The only known possible exception is the case of the offspring of living things, which contain inherited copies of the coded information for harnessing the energy and thus attaining the complexity of their ancestors. Anything less complex, including a recently deceased life form, will only degenerate no matter what natural conditions it may be exposed to.

In the face of this overwhelmingly obvious and universal observation, evolutionists can only point to slight increases in simple, static order, such as the formation of snowflakes or mineral crystals (which require a reduction in energy), or simple patterns of chaotically swirling masses, as seen in tornadoes, hurricanes, and boiling water. What a vast difference! How can it be considered scientific to have faith that the most complex sort of things known came into existence by simple natural processes, when there's no observation supporting the least trend toward such processes forming anything complex, but rather the universal observation of non-living things degenerating, living things that die especially, as they break down into their component elements? This faith shows the dictatorial grip that naturalistic philosophy has over this annexed area of science.

Then there's the case of the so-called "building blocks of life" (which is an analogy to something intelligently designed). These are relatively simple molecules that form things such as one half of one wrung of a DNA molecule, or one unit out of hundreds in a protein. DNA molecules and proteins are themselves but parts of the much more complex system of something as "simple" as a bacterium. Yet even at the level of the basic building blocks there are major problems with the belief that they could all form naturally and come together in a way that would support an organized, dynamic system, let alone become a living thing.

It took decades for intelligent scientists who were aiming at this specific goal to produce all of the amino acids and nucleotides needed for life (in what they optimistically considered recreations of past natural conditions). Even at that, it took separate experiments – no single experiment produced all the nucleotides together. Against this fact, evolutionists have only wishful stories about how the building blocks might have formed in different places, including outer space, and somehow happened to survive as they were moved by unguided forces of nature to some place on Earth where they could get together in just the right way, with nothing else interfering.

There are a number of problems that lessen the success of these experiments. In addition to producing only a subset of the building blocks of life, each experiment also produced a number of other amino acids and other chemicals which would have chemically prevented the required combination of the ones used in living things. News reports have generally ignored this problem, exulting in the production of any quantity of target molecules and almost never mentioning the great masses of gunk that would gum up the imagined life-forming process even if somehow all the needed building blocks could be produced and brought together.

Another problem not advertised by proponents in the popular press is that of chirality – the exact arrangement of the atoms in these molecules, which can be left- or right-handed. In living things, all the amino acids are left-handed, and all the sugars they use are right-handed. Outside of living things, including in the origin of life experiments, all these molecules naturally form in roughly equal numbers of the left- and right-handed forms! The evolutionists have to stretch even to grasp at straws in this case. For example, they cite the possibility of extreme conditions, although these have not been observed to produce more than a slight imbalance in the ratio. They can only hope that the imbalance could trigger some unknown process leading to the complete dominance of both left-handed amino acids and right-handed sugars.

Keep in mind that all this refers to just some basic units of living things! There are also great problems in getting them together, and adding all the other parts found in even the smallest living things. Also consider that ordinary things like water, heat, and agitation can cause these building blocks to break apart quickly, showing again that the evolutionists' stories about how life might have formed are unrealistic and wishful. They appeal to tiny traces of simple sugars and other simple molecules in space (or in meteorites) to support their faith that the more complex types must have existed in sufficient quantities and in the right place and conditions to play their role in the formation of life. They throw in conditions at volcanoes and the arctic and hot springs and hope that one of them could have been the lucky gathering place of all the right molecules. They point to simple bubbles of fatty oils and water, or even tiny holes in rocks, as the first containers of life or the molecules needed to form life, but the membranes of living things are filled with complex molecules designed to control the flow of different molecules (and even atoms) in and out at the right times and amounts. Evolutionists hope that some natural template such as mineral crystals would act as training wheels to help the first RNA or DNA start replicating, but even if it occurred, that would be a far different process from the complex series of interactions between large, intricate molecules that goes on in the simplest forms of life. DNA is so complex (in its coded information) and delicate that a complex system of other molecules is required to check it for damage and repair it.

It is ridiculous for evolutionists to wave their tiny straws of evidence and boldly call it sufficient support for faith in the natural origin of life when the most complex examples of intelligently-designed technology pale beside a simple bacterium. No man-made device is able to autonomously maintain itself in the natural world by gathering the raw matter and energy it needs, repair itself (including its system of encoding all the information it needs to do all this), and produce a complete copy of itself.

In short, the evolutionary stories of how life might have formed are like those of children desperately clinging to the belief that Santa Claus is physically real, denying all evidence to the contrary and excitedly clinging to anything that isn't. Or perhaps it's as if people refused to believe that an ancient city was designed and built by humans. They could note that the city was built of natural rocks or dried mud. What if it were carved out of rock, like the city of Petra? They could point to the natural formation of caves, and the formation of large, columnar blocks of lava, or the rough pattern of cracks in dried mud. They could point to shelters constructed by animals, as showing that inhabitable structures could be made without human intelligence. They might note the similarity of the buildings and claim it showed some natural force had reproduced them from a common template. It might become a very convincing story for those who felt some reason to believe it, or if the most prestigious and powerful people believed it. But no matter how many people believe in a story or how convincing it seems, if its supposed facts go against what everyone can see are truly undeniable facts, it should be rejected. The processes of nature that we have observed and know for sure are contrary to the idea that life arose and evolved by natural forces. Only faith, fantastic stories, and authoritarianism can make it seem otherwise.

Conclusion:

Beginning with Hutton, but especially after Darwin, scientists in geology and biology (and astronomy), in growing numbers turned their backs on the solid foundation of science: the study of how the world works, verified through repeatable observations and experimental demonstrations. Science, in some areas of these fields, became partly a game of Make Believe, guessing and telling stories about the past. The prime rule of this game is: you must explain everything in terms of natural forces alone. Since science only studied natural processes, this seemed like a logical extension, but none of the founders of science believed natural process could account for everything, and it is a gross departure from the scientific method.

At first the rules included the requirement that the forces could only be the ones known at the time, operating at the same rates observed in recent history, but that became too obviously impossible to believe. It should also be impossible to believe that life formed by natural processes alone, but that would allow the consideration of the intelligent input from a source outside the known universe, and that would call into question the foundation of this mode of conducting science.

It should also be very hard to believe that life began as the minimum possible form and developed into so many wonderful forms without any intelligent design, guidance, plan, or influence. Unfortunately, as in the story of the Emperor's New Clothes, the impossible has become the required belief. All the people with prestige and power in science, education, and the media refuse to consider that the many diverse forms of life may have required divine creation, or any sort of intelligent design. Neither can they tolerate consideration that creationism may be as reasonable and scientific a position as evolutionism. In many cases they have staked their reputations, their careers, and ultimately their eternal destinies on their faith that there is no God, or that God did not actively create the Universe and provide a clear account of doing so in the Bible.

Evolutionists know there is no way to demonstrate the truth of their claims about the ancient past, so they defend their faith with religious zeal, vigorously maintaining their monopoly on scientific institutions, public education, and the media. Through these they teach as facts things they know to be doubtful at best, inducting the next generation into their faith in the gods of Nature, Time, and Chance – the unholy trinity of Evolutionism. If a state or school should threaten to slip a tiny bit out of line, they call upon the power of the ACLU and the mockery of the mainstream media to join the authoritarianism of their expert witnesses to cow administrators, legislators, and judges, to punish the heretics, and to force even more stringent controls on the education system. Should any dissent arise in scientific institutions or academia, they do not hesitate to destroy people's careers and ruin their lives (as seen in the movie *Expelled: No Intelligence Allowed* and Dr. Jerry Bergman's book *The Slaughter of the Dissidents*).

Still, thousands of scientists, engineers, doctors, and others with advanced degrees in philosophy, medicine, engineering, and science continue to rebel against the authority of this evolutionary establishment. According to the evolutionists it is only because these rebels cling to fundamentalist religious beliefs. However, the evolution doubters include people of different religious beliefs and degrees of belief. Even some atheists have felt compelled to come up with alternative scenarios as far from the official story as their beliefs, careers, and reputations allow. Some of the Young-Earth Creationists were once atheistic or theistic evolutionists, or struggled with their religious beliefs, but then looked hard at the evidence and found the strongest of it went against evolutionism (see *Persuaded by the Evidence*, edited by Dr. Jerry Bergman and Doug Sharp). Evolutionists point out that many religious people of many denominations believe in evolution, but this only shows that if the evidence for evolution were truly strong, even fundamentalist creationists would find some way to reconcile their beliefs with evolutionism. No, people continue to point at evolutionism (AKA Evolution) and cry "The emperor has no clothes!" because there is nothing connecting the far-separated dots of data except the wispy threads of philosophy, faith, peer pressure, and authoritarianism.

Of course, there is much more to examine, mountains of details and data to learn and argue over. I wouldn't begin to think that this introduction alone will have convinced anyone to change their views to mine. There certainly is a lot of data that looks like strong evidence of billions of years. I'm not especially concerned about people believing as I do on the age of the Earth, although I think it is an important one. What I care about most is that people come to repent of their sinfulness and accept Jesus Christ as Lord and Savior. If you can do that and accept billions of years, and even evolutionism, everything else is secondary. Yet people often allow little things to keep them from the greatest good, and so in this book I have concentrated on the things of this world. I hope that it can be at least the start of a road to that wonderfully transforming relationship with the Son of God, through encouraging thinking that will overcome obstacles in the way that have been falsely presented as unquestionable historical and scientific facts.

Appendices

I had originally planned to include a good deal of extra discussion and data in four appendices, but I am now placing this information on my website, Fundamentalist Science (https://fundamentalistscience.com/)

List of Books (an informal bibliography)

Most of my education in evolutionary studies has been through the K-12th grade public school part of my academic training, plus many happy hours reading articles in popular magazines and technical journals such as *National Geographic, Science, Scientific American, Science News, Discover, Omni,* and especially *Nature.* Of course, as a Young Earth Creationist, I've read a lot more creationist material, including publications of all the major organizations, especially the *Creation Research Society Quarterly* (*CRSQ* for short). Here I list many of the books I've read that have also contributed to this book. I've only read the parts of a couple of the textbooks that were most relevant to evolutionism. I have marked in bold some which have special significance to me for one reason or another, and I recommend that you read them, too, even though they may not be directly related to creationism vs evolutionism.

In my conversations with people on both sides, I find it disappointing, or even annoying, to be confronted with rote responses and simplistic displays of knowledge with little if any indication of an intellect that is open, full of wonder, curiosity, and willingness to think outside of the box. The books in bold may help to foster such an intellect in you. For many people, reading anything in support of the other view might be quite a surprise; given the ubiquity of evolutionism in the media, I would guess this is more likely to be true of evolutionists, who have to make an effort against their usual inclination to even find a good example of creationist literature. Many also seem to be remarkably unaware of what Darwin wrote! On the other hand, many creationists are sadly unprepared for study of advanced evolutionary material. I hope this book will help both sides to be more understanding.

Format: Title, Author, publication data
***The Origin of Species*, Charles Darwin, Literary Classics, Inc., NY**

(includes appendix referring to "now" as 1861, and a 2nd edition of 1860)

Evidence as to Man's Place in Nature, T.H. Huxley, 1863, Intro by Selman Halabi, Barnes & Noble, N.Y, N.Y., 2006

Flatland: A Romance of Many Dimensions, Edwin A. Abbot, (published anonymously before 1883), Barnes & Noble 1963, Harper Collins 1983

Man from the Farthest Past, C.W. Bishop with C.G. Abbot & H. Hrdlicka, Volume 7 of the Smithsonian Scientific Series, 1930, 1934, Smith Institutional Series, Inc.

The New Evolution: ZOOGENESIS, Austin H. Clark, The Williams & Wilkins Co., 1930

The World of Copernicus, Angus Armitage, A Mentor Book by arrangement with Henry Schuman, Inc., 1947, 3rd printing 1953. Original title *Sun, Stand Thou Still,* published by The New American Library of World Literature, Inc,, NY, NY 1947

The Revolution in Physics: A Non-mathematical Survey of Quanta, Louis de Broglie, translated by Ralph W. Niemeyer, The Noonday Press, a subsidiary of Farrar, Straus & Cudahy 1953, 5th imprinting 1960

The Fossil Book: A Record of Prehistoric Life, P.V. Rich, T.H. Rich, M.A. Fenton &C.L. Fenton, Dover Publications, Mineola, NY, 1958, 86, 89, 1996

Asimov's Guide to Science, Isaac Asimov, Basic Books, Inc., NY, 2nd printing 1960, 1965, 1972

Monkey Trial: The State of Tennessee VS John Thomas Scopes, Sheldon Norman Grebstein, Editor, Houghton Mifflin, Boston, 1960

Creation or Evolution? David D. Riegle, Zondervan Publishing House, 1962, 1971

Science, God, and You, Enno Walthuis, Baker Book House Co., Grand

Rapids, MI,1963, 7th printing 1975

The Twilight of Evolution, Henry M. Morris, Baker Book House Co., Grand Rapids, MI, 1963, 15th printing 1974

Adventures with God: 15 Scientists Share Their Christian Faith, James C. Hefley, Zondervan Publishing House,1967, third printing 1970

Creation & the High School Student, Kenneth N. Taylor, Tyndale House Publishers, Wheaton, Illinois 1969

A Symposium on Creation, Donald W. Patten, et al., Bethany Fellowship, Inc., Minneapolis, Minnesota, copyright 1970 Baker Book House

The Creation of Life: a cybernetic approach to evolution, **A.E. Wilder-Smith**, 1st English edition 1970, 4th printing 1988, TWFT Publishers, Costa Mesa, CA

The World that Perished, John C. Whitcomb, Jr., Baker Book House Co., 1973

The Tao of Physics: An Exploration of the Parallels Between Modern Physics and Eastern Mysticism, Frijof Capra, Bantam Books, NY, Copyright Fritjof Capra 1975, 2nd Edition, Revised & Updated, 1984

The Dragons of Eden: Speculations on the Evolution of Human Intelligence, Carl Sagan, Ballantine Books, NY Copyright 1977 Carl Sagan, Ballantine Books edition 1978

Gödel, Escher, Bach: An Eternal Golden Braid, Douglas R. **Hofstadter**, Vintage Books, a division of Random House, NY, copyright 1979 by Basic Books, Vintage edition 1980

Tracking those Incredible Dinosaurs....And the People Who Knew Them, John D. Morris, Co-published by Bethany House Publishers and Master Books, a Division of CLP Publishers, 3rd printing 1984

He Who Thinks Has to Believe: A Thought-Provoking Allegory on the Origin of Life, **A.E. Wilder-Smith**, Bethany House Publishers, Minneapolis, 1981

Thermodynamics and the Development of Order, Emmet L. Williams, Editor, Creation Research Society, 1981, 3rd printing 1992

Chemical Evolution: An Examination of Current Ideas, **S.E. Aw,** Master Books, a division of CLP, San Diego, CA, 1982

Thread of Life: The Smithsonian Looks at Evolution, Roger Lewin, Smithsonian Books, Washington, D.C., Dist. By W.W. Norton & Co., NY, NY., 1982

What Is Creation Science (Revised & Expanded), Henry M. Morris, Gary E. Parker, 1982 Master Books, El Cajon, Ca., revised edition 1987

Design and Origins in Astronomy, **George Mulfinger, Jr.**, Editor, Creation Research Society, 1983

The Planiverse: Computer Contact with a 2-Dimensional World, **A.K. Dewdney**, Poseidon Press, NY, Pub. Pocket Books, a div. Of Simon & Schuster, 1984

Fossils that Speak Out: Evolution Vs. Creation, Startling Scientific Evidence, Phil Saint, P & R Publishing, Phillipsburg, NJ, 1985, 2nd edition 1993

Metamagical Themas: Questing for the Essence of Mind and Pattern, **Douglas R. Hofstadter**, Basic Books, NY, 1985

Quantum Reality: Beyond the New Physics, Nick Herbert, Anchor Press/Doubleday, Garden City, NY, 1985

The Riddle of the Dinosaur, John N. Wilford, 1985, A Borzoi book, published by Alfred A Knopf, NY, 1986

The Early Earth: An Introduction to Biblical Creationism, Revised Edition, John C. Whitcomb, Baker Book House, Grand Rapids, MI, 1972, Revised edition 1986, 6th printing 1992

The Revolution Against Evolution, Doug Sharp, 1986, Revised Edition 1993

The Scientific Alternative to Neo-Darwinian Evolutionary Theory, **A.E. Wilder-Smith**, The Word for Today Publishers, Costa Mesa, CA, 1987

Darwin's Enigma: Fossils and Other Problems, Luther D. Sunderland, Master Books, copyright 1988 Marilyn Sunderland,

The Emperor's New Mind: Concerning Computers, Minds, & the Laws of Physics, Roger Penrose, Oxford University Press, NY, 1989, reprint with corrections 1990

The Fingerprint of God: Recent Scientific Discoveries Reveal the Unmistakable Identity of the Creator, Hugh Ross, Promise Publishing Co., Orange, CA, 1989, 2nd edition revised & updated, 1991

The Long War Against God: The history and impact of the creation/evolution conflict, Henry Morris, Master Books edition, 1st printing, Copyright Henry Morris 2000, Original edition copyright 1989 by Baker Books

Beyond a World Divided: Human Values in the Brain-Mind Science of Roger Sperry, Erika Erdmann & David Stover, Shambhala, Boston, 1991

Darwin on Trial, Phillip E. Johnson, InterVarsity Press, Downers Grove, Illinois, 1991, 2nd edition 1992

Fossils: The Evolution and Extinction of Species, Niles Eldredge, Harry N. Abrams, Inc., N.Y. 1991

Bones of Contention: A Creationist Assessment of Human Fossils, Marvin Lubenow, Baker Books, Grand Rapids, MI 1992

Shattering the Myths of Darwinism, Richard Milton, Park Street Press, Rochester, Vermont, 1992, 1997

Creation Scientists Answer Their Critics, Duane T. Gish, Institute for Creation Research, El Cajon, CA - 1st edition 1993

The Biotic Message: Evolution Versus Message Theory, Walter J. Remine, St. Paul Science, Saint Paul, Minnesota 1993

The Dinosaur Society's Dinosaur Encyclopedia, Don Lessem & Donald F. Glut, Random House, NY 1993

The End of Physics: The Myth of a Unified Theory, David Lindley, Basic Books, a division of Harper Collins 1993

Creation: Facts of Life, Gary Parker, Master Books, Colorado Springs, CO 1994

Field Guide: Fossils, Richard Moody, Chancellor Press, Reed Consumer Books, Ltd., 1994

Mammoths, Adrian Lister & Paul Bahn, MacMillan, a Prentiss Hall Co., A Marshall Edition, Marshall Editions Development Ltd. 1994

The Evolution of a Creationist: A Layman's Guide to the Conflict between the Bible and Evolutionary theory, Jobe Martin, Biblical Discipleship Publishers, Rockwall, Texas, 1994, 2002, 2004

The Young Earth, John D. Morris, Creation-Life Publishers Inc., a division of Master Books, Colorado Springs, CO, 1994

After the Flood: The Early post-Flood History of Europe, Bill Cooper , New Wine Press 1995

Discovering Dinosaurs in the American Museum of Natural History, Mark Norell, Eugene Gaffney, & Lowell Dingus, A Peter N. Nevraumont Book, NY, NY, Alfred A. Knopf, Distributed by Random House, 1995

From Darkness to Light to Flight: Monarch – the Miracle Butterfly, Jules H. Poirier, Institute for Creation Research, El Cajon, CA 1995

Darwin's Black Box: The Biochemical Challenge to Evolution, Michael J. Behe, The Free Press, a division of Simon & Schuster, Inc., NY, NY 1996

Discovering Fossil Fishes, **John G. Maisey,** A Henry Holt Reference Book, Henry Holt & CO., NY, NY 1996

Noah's Ark: A Feasibility Study, John Woodmorappe, Institute for

Creation Research, El Cajon, CA 1996

Introduction to Physical Anthropology, Robert Jurmain, Harry Nelson, Lynn Kilgore, Wenda Trevathan, 7th edition, West/Wadsworth, an International Thompson Publishing Co. 1997

Not By Chance! Shattering the Modern Theory of Evolution, Lee Spetner, 1997, The Judaica Press, Inc., Brooklyn, NY, 1998

Buried Alive: The Startling, Untold Story About Neanderthal Man, Jack Cuozzo, 1998, Master Books, Green Forest, AR, fifth printing 2003

The Mythology of Modern Dating Methods: Why million/billion-year results are not credible, John woodmorappe, Institute for Creation Research, El Cajon, CA 1999

Tornado in a Junkyard: The Relentless Myth of Darwinism, James Perloff, Refuge Books, Arlington, Mass., 1999

Flatterland: like flatland, only more so, Ian Stewart, Perseus Publishing, copyright 2001 by Joat Enterprises

Search for the Truth: Changing the World with the Evidence for Creation, Bruce A. Malone, Search for the Truth Ministries, Midland, MI 2001

A Guide to Dinosaurs, Christopher A. Brochu, John Long, Colin McHenry, John O. Scanlon, Paul Willis, Cons. Ed. Michael K. Brett-Surman, Fog City Press, San Francisco, CA, 1997 US Weldon-Owen, Inc. Edition/Reprint 2002

Science Declares Our Universe Is Intelligently Designed, Robert A. Herrmann, Xulon Press, Fairfax, VA 2002

The Rise & Fall of Evolution: A Scientific Examination, Joseph A. Mastropaolo 2003
The Truth about Human Origins, Brad Harrub & Bert Thompson, Apologetics Press, Montgomery Alabama 2003

The Faces of Origins: A Historical Survey of the Underlying

Assumptions from the Early Church to Postmodernism, David Herbert, D & I Herbert Publishing, London, Ontario   2004

The Great Turning Point: The Church's Catastrophic Mistake on Geology – before Darwin, Terry Mortenson, Master Books, Inc. 2004

A Jealous God: Science's Crusade against Religion, Pamela S. Winnick, Nelson Current, a subsidiary of Thomas Nelson, Inc., Nashville, Tennessee 2005

Evolution and Other Fairy Tales, Larry Azar, Author House 2005

Genetic Entropy & the Mystery of the Genome, J.C. Sanford, Ivan Press, a division of Elim Publishing, Lima, NY   2005

The Politically Incorrect Guide to Science, Tom Bethell, Regnery Publishing, Inc., Washington, D.C. 2005

The Complete Book of Dinosaurs, Dougal Dixon, Hermes House, an imprint of Anness Pub. Ltd., London 2006

Creation, Fall, Restoration: A Biblical Theology of Creation, Andrew S. Kulikovsky, Mentor Imprint by Christian Focus Publishers, Ltd., Ross-Shire, Scotland 2009

The Mysterious Epigenome: What Lies Beyond DNA, Thomas E. Woodward & James P. Gills, Kregel Publications, a division of Kregel Inc., Grand Rapids, MI 2012

Flying Dinosaurs: How fearsome reptiles became birds, John Pickrell, NewSouth, Sydney, Australia 2014

Neither Darwin nor Deity: Making sense of being evolved, Shaun Johnston, Evolved Self Publishing, Tillson, NY 2016

Made in the USA
Columbia, SC
26 August 2018